COLLECTING ROCKS AND CRYSTALS

COLLECTING ROCKS AND CRYSTALS

Hold the treasures of the earth in the palm of your hand

John Farndon

Sterling Publishing Co., Inc.

New York

Library of Congress Cataloging-in-Publication-Data is available
upon request.

Published by Sterling Publishing Co, Inc.
387 Park Avenue South
New York, NY 10016-8810

Distributed in Canada by Sterling Publishing
c/o Canadian Manda Group
One Atlantic Avenue, Suite 105
Toronto, Ontario, Canada M6K 3E7

This book was designed and produced by
Quarto Publishing plc
6 Blundell Street
London N7 9BH

Project Editor Marnie Haslam
Editor Angela Koo
Art Editor Julie Francis
Designer Liz Brown
Photographer Les Weis
Illustrator Kuo Kang Chen
Art Director Moira Clinch
Picture Researchers Gill Metcalfe, Marnie Haslam
QUAR.R&CC

Manufactured in Hong Kong by Regent Publishing Services Ltd
Printed by Winner Offset Printing Factory Ltd

ISBN 0-8069-3147-7

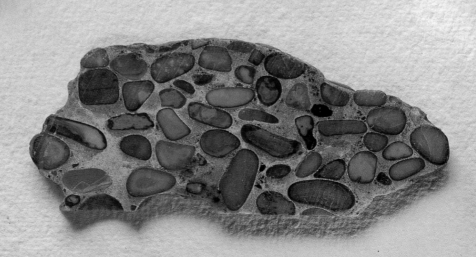

CONTENTS

Introducing Rocks and Crystals

Rocks and crystals are the raw materials of the Earth's surface—the material beneath every hill and valley, mountain and plain. Some are just a few million years old. Others are almost as old as the Earth.

What are rocks?

Rocks are never far beneath the ground. They are only exposed on the surface in a few places—such as bare rock outcrops, cliff faces and quarries. But dig down almost anywhere on the Earth's surface and you will come to solid rock before long.

Like the other smaller planets in the solar system, our world is made almost entirely from rock. The Earth is a bit like a perfectly boiled egg—with a semi-liquid yolk or "core," surrounded by a thick, soft layer called the mantle, and covered by a thin hard shell called the crust. The core in the very center is metal but the crust and mantel are made entirely from rock.

ROCKS AND MANKIND

No wonder, then, that rocks have played such an important part in mankind's history. Rocks were used by humans for their very first cutting tools, millions of years ago. At least three million years ago, early hominids (manlike creatures) were chipping the edge off hand-sized round pebbles, perhaps to use as weapons. Two million years ago, hominids began using flints to make two-sided hand-axes, which is why the first age of man is known as the Stone Age.

Later, clay was used to make pottery, and since then mankind has found an increasing variety of ways to use rocks. They can be broken up and reshaped to provide building materials for everything from cottages to cathedrals, harbor walls to roads. Certain minerals—the natural chemicals they are made from—can be extracted or processed to make a huge range of materials. All metals, such as iron, copper, and tin, come from minerals contained in rock. So do most

ROCK SOURCE (below)
One of the best places to see living rock is in quarries, where it is blasted and dug from the ground to provide building stone and other materials.

ROCK OUTCROP
(right) Another good place to see rocks is on the coast where the relentless pounding of the waves exposes them in cliffs and rocky platforms. Beaches are made from ground-down fragments of rock. The rock shown in this outcrop is chalk.

of our fuels, such as oil, coal, and natural gas, as well as the sand we use to make glass and much more.

We will never be able to use up even a tiny fraction of all the rocks on Earth, but particular kinds of rocks—and many of the minerals they contain—are much rarer than others and much harder to find. Moreover, our consumption of certain minerals has increased so rapidly that more of them have been extracted from the ground in the last few decades than in all of Earth's history. Rocks and minerals are, like the living environment, precious resources, and should not be squandered.

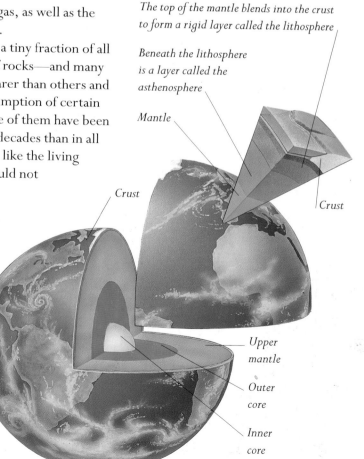

The top of the mantle blends into the crust to form a rigid layer called the lithosphere

Beneath the lithosphere is a layer called the asthenosphere

Mantle

Crust

Crust

INSIDE THE EARTH *(right) Vibrations from earthquakes have revealed that the Earth is made mostly of rock—but its interior is quite complex. We stand on a thin shell or "crust" of solid rock, little more than 40 miles thick at most. Beneath the crust is a thick mantle of rock so hot it flows like maple syrup, but very slowly. At the heart of the earth is a core of metal (iron and nickel). The outer core is always molten, but pressure in the inner core is so great it remains solid, even though temperatures here reach 7000°C.*

Upper mantle

Outer core

Inner core

What are rocks made of?

EVERYONE KNOWS that rock is a hard, generally very tough, solid material, usually strong enough to use for building or to support great weights. But other solids can be used like this, too. What makes rock unique is that it occurs naturally—and that it is made up from tiny crystals joined together.

CRYSTALS

Crystals are a very particular kind of solid. What makes them distinctive is that they are regular, geometrical shapes, with smooth faces and sharp edges and corners—chunky rather than globby. Many, though not all, are shiny or clear to look at. Indeed, they got their name from the clear, glassy chunks of quartz which the Ancient Greeks called *krystallos* because they believed it was unmeltable ice. Modern scientific analysis has shown that crystals are made up of atoms that link together to form a regular structure or lattice.

Some of the crystals that make up rock are so small that they are invisible to the naked eye, and even a cubic inch of rock contains many thousands of them. A few rare crystals are as big as telephone poles, and they are found amongst smaller crystals like a sausage in a tin of baked beans.

MINERALS

All crystals in rock are made from natural chemicals called minerals. All but a few minerals form as crystals so, in this book, whenever a mineral is referred to, you can assume it is crystalline (crystal form) unless otherwise specified. Definitions of what a mineral is vary, but most geologists tend to agree that they are usually naturally occurring inorganic (non-living) solids with particular chemical compositions and crystal structures. A few rocks are made from crystals of just a single mineral; many are made from half a dozen or more.

GRANITE (below) Look at any rock closely enough, and you will see that, like this granite, it is not smooth, but made from tiny grains or crystals. All these crystals are made from natural chemicals called minerals.

CARVED STONE (left) Sandstone is made of countless grains of sand—tiny crystals of the tough mineral quartz. Grains can easily be cut away to create the desired shape—but the carved surface withstands the worst weather for centuries, because the individual grains are so durable.

Most minerals are compounds of two or more chemical elements, although a few such as gold and copper, called native elements, are made of an individual element. There are well over two thousand minerals, each made from a particular combination of chemicals, but only 30 or so are very widespread. Most are present in rocks only in minute traces, and are easy to spot only when they become concentrated in certain places by geological processes. These concentrations give us the ores from which many metals are extracted, and the brilliant gems which give us jewels.

HOW MINERAL CRYSTALS FORM

Crystals form when liquid evaporates or molten solids cool, and chemicals dissolved within them solidify. They actually grow as more and more atoms attach themselves to the structure—just as icicles grow as more and more water freezes on them.

Some crystals form as hot, molten rock from the Earth's interior slowly cools. Some form from chemicals dissolved in watery liquids within the ground. Some are formed as minerals are altered chemically. And some form as other crystals are squeezed or heated so hard by geological processes in the ground that they form entirely new crystals.

MARBLE *(below) In most rocks, crystals form when minerals solidify from water or molten rock. In rocks like marble, they re-form from melted crystals.*

CRYSTAL MASS *(below) A microscope photograph reveals the mesh of crystals a rock is made from. This rock is syenite, an igneous rock (see page 12) made mainly from crystals of the minerals feldspar and quartz (pages 74–75).*

GROWING CRYSTALS *(above) Crystals slowly grow as particles of minerals dissolved in liquid link together and solidify. Suspend a grain of sugar in very, very sugary water and watch it grow over several weeks.*

Crystal systems and shapes

THE CRYSTALS you find in the ground often look chunky, but are rarely quite the neat and perfectly regular shapes you see in drawings. Nevertheless, by careful study, crystallographers have come to realize that all crystals are made up in certain ways, and form a limited range of shapes.

The crystals that form at the same time as the rock are usually small. All the different crystals are mixed up together and the characteristic shapes are much harder to spot. However, in certain places (see page 15) larger individual crystals and clusters of particular kinds of crystals form later than the rock and so the characteristics are much easier to recognize.

(see page 15)

CRYSTAL SYSTEMS
(below) In ideal conditions, crystals grow in remarkably regular shapes. They are essentially symmetrical, and every crystal can be classified into one of six groups or "systems" according to how sides and corners are symmetrical.

SYMMETRY

Individual crystals are all essentially symmetrical—that is, they look the same shape from a number of different angles—though the symmetry is often far from perfect in nature. Every crystal has an axis of symmetry, which is an imaginary line drawn through its center in a certain direction. If you turn the crystal about its axis of symmetry, it always appears perfectly symmetrical. Cubes are very symmetrical indeed, and you can draw an axis of symmetry through one in twelve different directions. Most shapes are less symmetrical and have fewer axes.

CUBIC *Of all the systems, the cubic is the most perfectly symmetrical.*

MONOCLINIC *crystals are essentially symmetrical in just one plane.*

HEXAGONAL/TRIGONAL *crystals both have six symmetrical sides.*

TRICLINIC *With three slightly inclined axes of symmetry, the symmetry of these crystals is hard to spot.*

TETRAGONAL *crystals look like tall boxes, with a square cross-section, but corners may be shaved off.*

ORTHORHOMBIC *crystals have three axes of symmetry at 90°—but each is a different length.*

CRYSTAL SYSTEMS

Every individual crystal can be classified according to its symmetry into one of six different systems. These six systems are named after the words for six perfect geometrical shapes: monoclinic, triclinic, cubic, tetragonal, orthorhombic, and hexagonal or trigonal.

Crystals of the same mineral always fit into the same system, so identifying the system is an important clue to identifying the mineral. But it is not always easy to be sure which system it fits into, because crystals within each system can take a number of different forms. Cubic crystals, for instance, may be cubes, but they can also be octahedrons (eight-sided shapes) and various other shapes. What all these forms of cubic crystal will have in common, though, is that they will have the same twelve axes of symmetry as a cube.

ARAGONITE *(above) Like topaz, aragonite crystals are orthorhombic, but the crystals are typically tabular (long and flat) and grow in long needles.*

CRYSTAL HABITS

Individual crystals tend to grow together in a mass. When they form at the same time as the rock, every crystal of the same type tends to grow together in a particular way. So each kind of crystal forms a mass with a characteristic shape or "habit"—for example, dendritic (tree-like) or acicular (needle-shaped). On the whole, each kind of crystal has a particular habit because it tends to form in certain conditions, so you can expect to find particular habits in particular places.

TOPAZ *(right) The symmetrical form of crystals can be seen most clearly in large crystals that grew slowly, like this topaz. Topaz has "orthorhombic" symmetry, and typically forms in a shape that looks a little like a prism.*

How Rocks are Made

ROCKS COME in many shapes, textures, and colors, but they all form in one of three ways. Igneous rocks form from the molten material which comes from the Earth's hot interior. Sedimentary rocks form from compressed layers of debris settling on places like the seabed. Metamorphic rocks form as other rocks are transformed by the huge heat and pressure of intense geological activity.

Igneous rocks

THE EARTH IS not as solid as it seems. Just a few thousand feet below the Earth's surface it is very hot. A few dozen miles below the surface, the mantle reaches a searing 1,400°C. This is hot enough to make the rock so soft that it flows like maple syrup. Pockets of hot mantle rock can melt and bubble up through the Earth's crust. Molten mantle rock, or "magma," is incredibly hot, but as it nears the surface, it may start to cool. As the magma solidifies, crystals grow within it, until it becomes a solid mass of hard crystalline rock called igneous rock.

GUSHING LAVA *(above)*
Extrusive igneous rocks form mainly from molten lava, gushing onto the Earth's surface in volcanoes.

Sometimes magma spews out onto the surface through volcanoes before it cools and solidifies. Rock formed like this is called "extrusive" igneous rock. Sometimes it cools before it reaches the surface, forming masses of "intrusive" igneous rock below the ground.

Igneous intrusions may exist as large masses called batholiths—formed as magma arches up into the ground—or thin sheets called dikes and sills. Concordant intrusions are found in existing cracks in the rock, while discordant intrusions burst up through the ground so forcefully that they open up new cracks which cut across existing structures.

Concordant intrusions flow into existing structures within the rock.

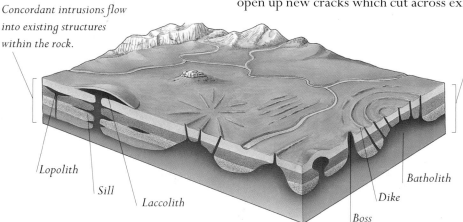

Discordant intrusions burst their way through the rock.

TYPES OF INTRUSION *(left)*
Molten magma solidifies beneath the ground into igneous intrusions in dozens of different forms.

Lopolith

Sill

Laccolith

Boss

Dike

Batholith

Sedimentary rocks

ALMOST 90 PER CENT of the Earth's crust is made up of igneous rock, but three-quarters of the continents are covered with thin layers of debris—most of it the fragments of igneous rock broken down by exposure to the weather. This debris settles on the beds of oceans, lakes, and rivers, or is piled up by moving sheets of ice or the wind in deserts. As debris builds up over millions of years, the layers are gradually compacted into solid sedimentary rock. Layers of mud can be compacted into as little as a tenth of their original thickness.

As sediments are compacted, they are usually glued together by cements made from materials dissolved in the water from which the debris settled. The most common cements are calcite (page 72), silica (page 74), and iron compounds, which give rock a red, rusty look.

The original layers in which the debris settle are often clearly visible in sedimentary rocks as lines called bedding planes. Where the rock has been undisturbed, these lines are horizontal, but often movements of the Earth's crust twist them into all kinds of contorted shapes. There may also be vertical cracks, called joints, which form as the rock dries out.

RIVER-BORN (above)
Most sedimentary rocks form from millions of rock fragments washed down to the sea by rivers.

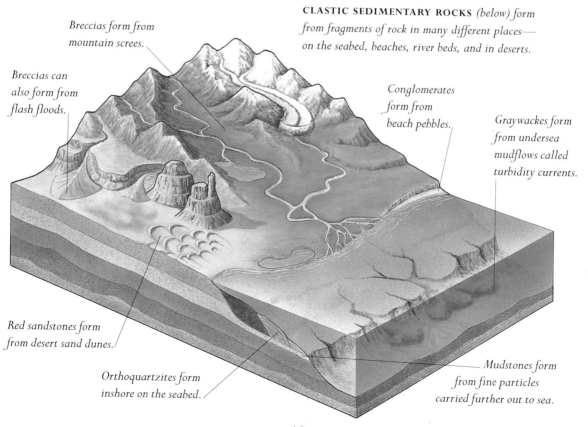

Breccias form from mountain screes.

Breccias can also form from flash floods.

CLASTIC SEDIMENTARY ROCKS (below) form from fragments of rock in many different places— on the seabed, beaches, river beds, and in deserts.

Conglomerates form from beach pebbles.

Graywackes form from undersea mudflows called turbidity currents.

Red sandstones form from desert sand dunes.

Orthoquartzites form inshore on the seabed.

Mudstones form from fine particles carried further out to sea.

Metamorphic rocks

WHEN ROCKS ARE seared by the heat of molten magma or crushed by the huge movements of the Earth's surface that build mountains, they can be altered beyond recognition. The crystals they are made from re-form so completely that they become, in effect, new rocks, called metamorphic rocks—after the Greek word *metamorphosis*, which means "transformation." Igneous, sedimentary, and metamorphic rocks can all be metamorphosed into new metamorphic rock.

Rocks can be altered by heat if they are close to an igneous intrusion or are deep underground. They can be altered by pressure when they are buried more deeply by overlying material or crushed if they are in a zone where the great "tectonic" (see page 76) slabs of rock that make up the Earth's surface are moving together.

Heat and pressure alters rocks in two ways. First, it changes the mineral content, by making minerals react together to form new ones. Secondly, it changes the size, shape, and alignment of the crystals, breaking down old crystals and forming new ones in a process called recrystallization.

METAMORPHIC ROCK *(above) Schist is one of the most common metamorphic rocks and forms where other rocks are subjected to heat and pressure under mountain ranges.*

CONTACT METAMORPHISM *(below) is when rocks are transformed by direct contact with the extreme heat of an igneous intrusion.*

GNEISS *(above) is the metamorphic rock that forms when crystals are subjected to the most extreme heat and pressure.*

REGIONAL METAMORPHISM *(right) is the alteration of rocks over a large area as the Earth's surface moves, creating intense heat and pressure—often throwing up huge mountain ranges.*

Sandstone is altered to metaquartzite.

Limestone is altered to marble.

Fine sediments are altered to hornfels and spotted rock.

Hot magma forms batholith

Slate forms from shale under mild heat and pressure.

Schist forms under medium heat and pressure.

Gneiss forms under intense heat and pressure.

How good crystals form

ROCKS CAN BE found almost anywhere, and all rocks are made from mineral crystals. One way crystals can form is as a result of a rich mixtures—when magma cools and interacts with pieces of native rock, thereby changing the magma's chemical composition.

Another way crystals can form is as a result of meteoric water—when hot magma heats groundwater up. The combined force of pressure and heat enriches the water with all kinds of dissolved elements, such as sulphur, mercury, or copper. Washed through cracks to cooler, less pressured places, these dissolved minerals solidify over thousands of years to form valuable crystal deposits. Crystals can also form when hot magma packed with minerals seeps upward through the layers of the Earth in the form of a vein. Minerals with the highest melting points crystallize first, so the chemical mix of the molten magma gradually changes throughout the different layers. Gold and silver metal deposits are commonly formed in this way.

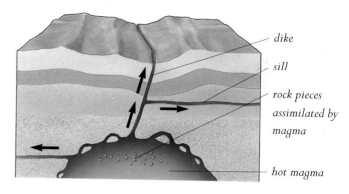

dike

sill

rock pieces assimilated by magma

hot magma

RICH MIXTURES (above) *Magma can give birth to many kinds of crystals as it cools and interacts with the native rock in many ways, each subtly changing the magma's chemical composition.*

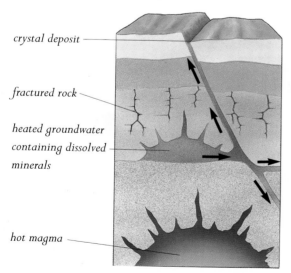

crystal deposit

fractured rock

heated groundwater containing dissolved minerals

hot magma

METEORIC WATER (above) *Many good crystals form when water is heated by magma underground and becomes enriched with minerals.*

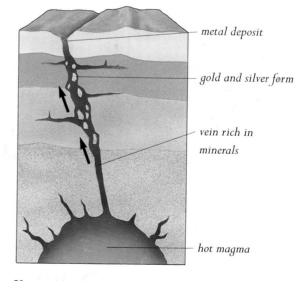

metal deposit

gold and silver form

vein rich in minerals

hot magma

VEINS (above) *rich in minerals such as gold and silver are formed when hot, watery liquids made from magma underground seep upward through cracks.*

Rocks in the Landscape

THE MAJORITY OF rocks on the surface of continents are sedimentary, and each formed at a particular time in Earth's history. By looking at the layers, or "strata," geologists can tell an enormous amount about how and when they were formed—and what has happened to them since.

When did the rocks form?

BACK IN THE seventeenth century, a Danish geologist called Nicolas Steno (1638–1686) stumbled onto something very important about sedimentary rocks. He realized that layers of sediment must be laid down one on top of the other. If so, then, the oldest layers are at the bottom and the youngest must be at the top. This is called the law of superposition.

The law of superposition works as long as the rock layers have not been overturned or contorted since they were formed. As soon as layers are broken up or separated from each other, it becomes much harder to tell which is the oldest and which the youngest. Moreover, very different kinds of rock were laid down at the same time in different places. So how do geologists tell how old different rocks are—or if they are the same age?

The answer is, essentially, fossils—the petrified remains of living plants and creatures preserved in the rocks as they formed. In the early nineteenth century, an English engineer called William Smith (1769–1839) noticed that each layer of rock in a sequence

LAVA BEDS (above)
Sedimentary rocks form in a definite sequence, with the youngest layers on top—but the ferocity and chaos of volcanic activity means lava of all ages tends to be mixed up in igneous rocks.

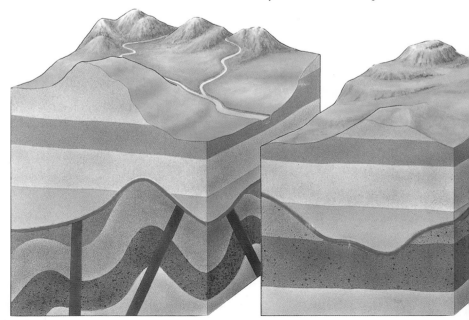

Angular unconformity, where one sequence cuts across another at an angle.

Disconformity, where a bumpy line gives away the break in the sequence.

contains a certain range or "assemblage" of fossils—and that the same range is repeated in different rocks. Smith realized that if two layers of rock have the same set of fossils, they must be the same age, even if one is sandstone and the other clay. So rocks can be dated simply from the range of fossils they contain. Other methods are used, such as radioactive dating, but fossils remain the key to "stratigraphy"—the study of the distribution and order of rock layers.

BREAKS IN THE SEQUENCE

Although, generally speaking, sediments are laid one on top of the other in a particular sequence, every now and then there is a dramatic break—and one sequence of rock can appear directly on top of a completely different sequence. This break is called an unconformity. This can happen, for instance, when a folded sequence of rock is worn down by erosion to give a flat surface cutting across all the layers—and is then buried beneath a new sequence of rock strata. When this happens, the unconformity is called an angular unconformity, because the lower, older strata tilt at a very different angle to the younger strata above.

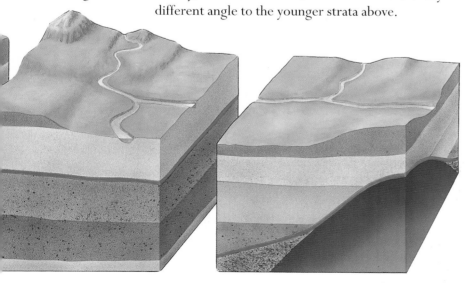

Parallel unconformity, where only a great age gap gives away the break.

Non-conformity, where a sedimentary sequence overlies igneous or metamorphic rock.

NATURAL BREAKS
(left) Unconformities are breaks in the normal sequence of rocks, where rock from one sequence lies over rocks from a different sequence. The red line in each of these four pictures indicates the various ways these natural breaks can happen.

Using fossils

IN NEARLY EVERY sedimentary rock, there are the preserved remains of plants and animals that lived millions of years ago when the rocks were formed. These fossils can be the geologist's most useful clues to the history of rocks.

When an animal dies, its soft parts rot quickly away, but if its shell or bones are buried swiftly, they may eventually turn to stone. Most fossils are either shells or a few isolated bones. Complete skeletons are rare, and the soft parts of a body are almost never preserved. But a knowledge of anatomy gained from living creatures allows paleontologists (who study fossil life-forms) to learn about the creature.

The vast majority of fossils are shellfish that lived in shallow seas. Fossils of soft-bodied creatures such as land-living mammals are rare, because they rotted away before they could be fossilized. So when geologists search for fossils to date rocks, they are generally looking for fossil shells such as those of brachiopods, cephalopods, and trilobites.

BRACHIOPOD (above)
Brachiopods are now rare shellfish with broad, flat vivalve shells

MONOGRAPTUS
(above) Monograptus was a tiny, worm-like creature that lived in Silurian Times.

ECHINOID (left)
Echinoids are basically sea urchins.

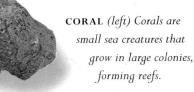

CORAL (left) *Corals are small sea creatures that grow in large colonies, forming reefs.*

Dead shellfish — Soft body parts quickly rot away — Water

1

2 — Hard shell is buried in sediment — Soft sediment

3 — **5** — Preserved unaltered — Compacted sediment

Hollow mould

4 — **6** — Destroyed by pressure and heat — Metamorphic rock

A replica is formed

FOSSIL FORMATION *(above) When a creature such as a shellfish dies and falls to the sea floor, its soft body parts quickly rot away (1). But the hard shell, made mainly of the mineral aragonite (calcium carbonate), may be buried intact by sediment (2). Over millions of years, the shell may be dissolved away by water trickling through the sediments, sometimes leaving an empty cast or mould (3). But the dissolved aragonite often recrystallizes within the cavity like jelly in a mould, leaving a perfect replica or "fossil" of the shell in calcite (4). Occasionally, animal remains may be preserved hard but virtually unaltered (5). The shell is destroyed when sedimentary rock is metamorphosed.*

INDEX FOSSILS

A fossil is little use for dating rocks if: it occurs in only a few rocks, if it is hard to identify, or if it has changed so little over time that it appears in rocks of all ages. So geologists look for key kinds of fossils, which they call index fossils. If they spot a particular index fossil in a rock stratum, it helps them pinpoint the age of the rock very quickly.

For a fossil to be good as an index, it must be widely distributed and easy to recognize. It must also be small and have changed rapidly through time, showing marked changes at different points. Among the best index fossils for dating rocks worldwide are shellfish that floated widely throughout the oceans such as ammonites (a kind of cephalopod) for the Jurassic and Cretaceous Periods (pages 20–21), goniatite ammonites for Devonian, Pennsylvanian, Mississippian, and Permian, and graptolites for the Ordovician and Silurian Periods. Crinoids (sea lilies) and trilobites are good index fossils for the Cambrian. On a local basis, brachiopods, cephalopods called belemnites, and corals make useful index fossils.

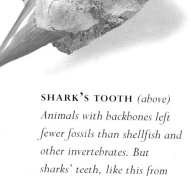

SHARK'S TOOTH (*above*)
Animals with backbones left fewer fossils than shellfish and other invertebrates. But sharks' teeth, like this from the Eocene, a type of mako shark, are quite common.

BIVALVE (*above*)
Bivalves are shellfish, such as mussels and clams, with a shell of two hinged halves.

CEPHALOPOD (*right*)
Cephalopods are squid-like shellfish, like the extinct ammonite.

TRILOBITE (*above*)
Trilobites are sea creatures with a three-part flexible shell that were common 350 million years ago but are now extinct.

The geological column

IF ROCK SEDIMENTS remained forever undisturbed, you could slice right down through them to reveal the whole sequence of rocks that have formed on Earth, from the youngest at the top to the oldest at the bottom. In fact, if you could extract a column of rock right down through the sequence, it would be like a book telling the complete story of Earth's history. If sediments had been deposited continuously since Cambrian times, the column would now be 100 miles deep.

Although such a column cannot exist anywhere on Earth, this "geological column" is a very useful idea, and it is the standard way of showing the history of sedimentary rocks. The column only goes back to Cambrian times, some 570 million years ago, because it was only after the Cambrian that shelly and bony life-forms became widespread enough to leave a good fossil record. Very little is known about the four billion years of Earth history before that time—known as Pre-Cambrian time.

HISTORY IN ROCKS (below) From the sequences of rock laid down in different parts of the world, we can piece together the entire story of life on Earth, and have a snapshot glimpse into all the eras of geological history.

1 PRECAMBRIAN TIME 4600 MYA
The first single-celled life forms developed, followed by algae which slowly added oxygen to our atmosphere. Later in the period, multi-cellular soft-bodied organisms such as worms and jellyfish appear.

2 CAMBRIAN PERIOD 570 MYA
No life on land. But a wealth of small invertebrates, such as shellfish appear in the seas.

3 ORDOVICIAN PERIOD 510 MYA
Crustaceans (like crabs) appear, and early marine animals, and coral-reefs.

4 SILURIAN PERIOD 438 MYA
Fish with jaws appear and some fish live in rivers and lakes. The first plants appear on land.

5 DEVONIAN PERIOD 410 MYA
Sharks in the sea. Insects on land, and then the first amphibians like Ichthyostega. Vast forests of giant ferns and mosses.

6 CARBONIFEROUS PERIOD 355 MYA
Vast swampy forests which eventually form coal deposits. The first reptiles.

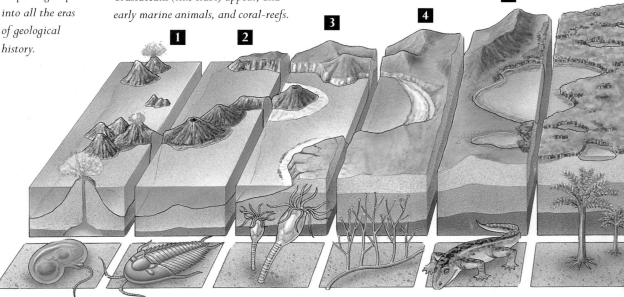

Blue-green algae (single-celled life) — Trilobite (small invertebrate) — Crinoid (marine animal) — Cooksonia (land plant) — Ichthyostega (amphibian) — Giant tree fern

Just as a day is divided into hours, minutes, and seconds, so geological time is divided into units called chronomeres. These are times when particular events or processes appeared in Earth's history. The longest chronomeres are Eons. Eons are divided into Eras, Eras into Periods, Periods into Epochs, Epochs into Ages and Ages into Chrons. For geologists, the most important divisions are Periods, which last around 30–60 million years, such as the Devonian and Silurian.

MYA means Millions of Years Ago.

7 PERMION PERIOD 290 MYA
The first conifers appear. Reptiles flourish, though deserts are widespread and many creatures die out.

8 TRIASSIC PERIOD 250 MYA
Small mammals and marine reptiles are seen for the first time. Seed-bearing plants dominate.

9 JURASSIC PERIOD 205 MYA
The age of dinosaurs begins. Bird-like reptiles appear.

10 CRETACEOUS PERIOD 135 MYA
Flowering plants and small land mammals appear. Oil and gas deposits form. Dinosaurs die out at the end of the Period.

11 TERTIARY PERIOD 66 MYA
The first large mammals as grasslands expand and the first modern birds appear. Primates evolve.

12 QUATERNARY PERIOD 1.6 MYA
Many mammals die out in the Ice Ages. Modern humans (homo sapiens) appear.

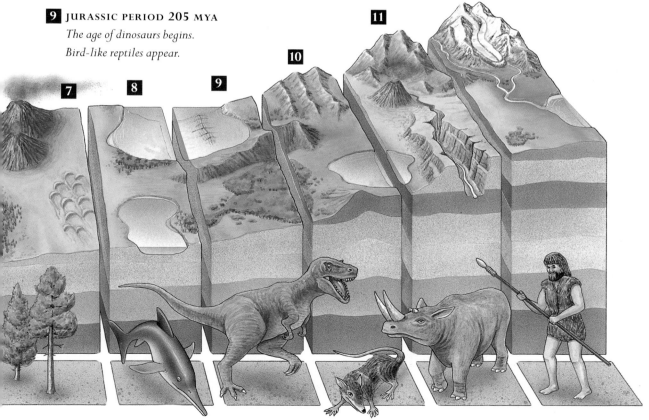

Conifer   Mixosaurus (marine reptile)   Tyrannosaurus (dinosaur)   Crusafonia (small mammal)   Arsinoitherium (large mammal)   Homo sapien (modern human)

Building Your Collection

Y OU DON'T NECESSARILY need any equipment at all to start building a rock and crystal collection. You can simply keep an eye out for loose specimens and pick them up when you are out walking in the countryside or by the sea. But getting a few basic tools will help you get the best samples.

What you need

B Y FAR THE MOST important tool is a hammer for breaking up loose pieces of rock. You can get away with an ordinary bricklayer's hammer and a chisel, but it is much better to invest in a proper geological hammer. These typically have a small tapered head and a tail drawn out in a long point which is good for levering samples out. A chisel is a useful addition.

    The geologist's rule is to hammer as little as possible. Hammering encourages erosion and leaves scars on the rock face. In fact, you should not really use a hammer for knocking samples out of the living rock at all—only for breaking rocks up on the ground. Rocks can splinter when hit, so it is vital to wear goggles when hammering them. Also essential for protection are strong gloves and, if you are visiting cliffs or quarries, a hard hat as

ROCK HUNTING KIT

(below) These are the most important items of equipment you will need when going hunting for rocks and crystals.

hammer   goggles   gloves   hard hat

protection for the head. Another useful item is a magnifying glass. This helps you both to spot small crystals in the field and to identify crystals you have found. Don't make the mistake of getting too powerful a lens. A twenty-times magnification lens shows too small an area to be useful in the field. The best magnification is about eight to ten times.

A multi-purpose penknife and a small shovel can sometimes prove invaluable.

You will also need a strong shoulder bag of some kind to carry samples in. A small rucksack leaves your hands completely free to pick up samples. Fill it with some bubble-wrap or old paper to wrap the samples and to stop them knocking into each other.

Finally, you need maps (and maybe a compass) for helping you find your way around and for locating good sites. You should also have a notebook for recording where you made your finds, and pens and sticky labels for labelling them on the spot. A camera to record the site is also useful—and sometimes saves digging out a sample unnecessarily.

bubble wrap

penknife

compass

notebook and pen

magnifying glass

maps

Understanding geological maps

HOW TO MAKE A
CROSS-SECTION OF A
GEOLOGICAL MAP

A**N ORDINARY MAP** can show a great deal once you know what to look for, but a geological map is even more useful. Geological maps show the types of rocks that occur in a particular place. With a good geological map, you can work out exactly where you are likely to find good samples. An experienced geologist can also use it to work out the structure of the rock below the ground, how different strata relate, and even how the different rock formations were shaped through time.

SOLID AND DRIFT MAPS

There are two main kinds of map. Most are "solid" geology maps, which means they show the pattern of solid rocks immediately beneath the surface. These will give you the best idea of what kinds of rocks and crystals you are likely to see exposed in rock faces, cliffs, and quarries. The other kinds are "drift" maps, which show what loose sediment there is on the surface, including the gravels deposited by rivers and glaciers. These sediments are the fragments of solid rocks broken down over the ages, and so drift maps may help you locate useful river placer deposits (see page 27).

Geological maps usually show the different rocks with patches of different colors. Igneous rocks are typically shown in shades of crimson, purple, and magenta. Sedimentary rocks are in dull browns, yellows, and greens—except for limestone which is usually blue-gray. Metamorphic rocks are in shades of pink and gray-green.

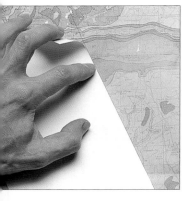

STEP 1

*Lay a straight edge of
paper where you want your
section—typically at
right angles to the main
contour lines.*

STEP 2

*Mark off where each
contour line and each rock
band (in different colors)
crosses the edge.*

STEP 3

*Lay the edge on a piece of graph paper
and make a dot above each of your
marks at the appropriate height on the
graph. Join the dots.*

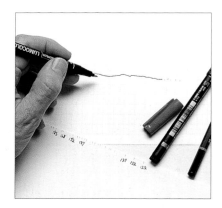

STEP 4

*Mark on the graph where the rock bands
occur. Now see if you can guess the
structure of the underlying rock beds.*

OUTCROPS AND SECTIONS

What solid geological maps also show is rock outcrops—that is, places where a solid rock formation reaches the surface. Usually, outcrops are covered by a thin layer of soil or other sediments. But occasionally they are completely bare, in which case they are called exposures. It is these exposures—such as cliff faces, quarries, and tors—that are the best sites to look for samples. They are typically marked by hatching on the map.

One way of really understanding the way rocks are arranged in the ground is to look at the cross-section which often appears on the bottom of a map. This is a vertical slice through the rock that shows the different layers. The sections on maps are usually worked out with the aid of seismic (vibration) surveys and holes drilled in the ground. But an experienced geologist can work out a rough cross-section from the pattern of rocks on the surface and the shape of the landscape.

GEOLOGICAL MAP *(below right)*
The Grand canyon is one of the most fascinating geological sites in the world. The Colorado river has cut down through the Plateau over millions of years to reveal a complete sequence of rock beds. Overlaying the solid beds, however, are recent river and mudflow deposits. These are shown clearly on this drift map of the area. The multitude of different deposits and deposits of different ages are shown here by many different colors.

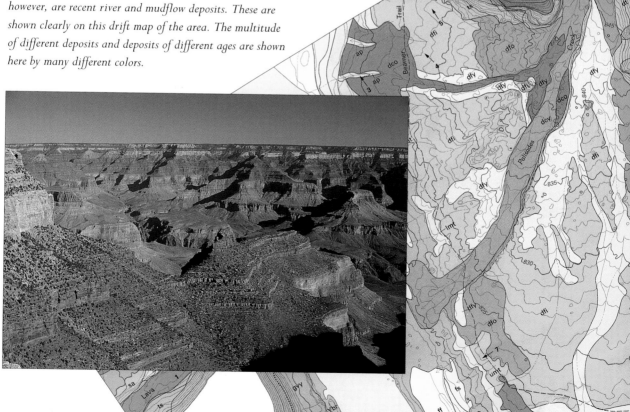

Where to find rocks and crystals

Y OU CAN SEE rocks everywhere. Houses may have sandstone walls and slate roofs. Office blocks may have granite walls. Statues may be made of marble. The most precious jewels are usually mineral crystals. And, of course, geological museums and rock shops are full of good samples. But if you want to build up your own collection, you need to go out into the countryside hunting.

There are two kinds of sites in which you can find rock and crystal samples: rock outcrops and deposits. Rock outcrops include exposed hilltops, cliffs, quarries, cuttings, and building sites. Deposits include all the places where rocks and crystals are found lying loose, such as river beds, beaches, fields, and yards.

When hunting for rock, you have to remember that nearly all land belongs to someone, and you may need their permission to collect samples. Indeed, any samples found on their land actually belong to them. Seeking permission is especially important if you want to hunt in old quarries. You should also check whether the site is part of a Nature Reserve or National Park where hammering or collecting samples is forbidden. The rule when rock and crystal

COASTAL CLIFFS *(above)*
One good place to hunt for rocks is at coastal cliffs, where the sea exposes layers of rock and pounds them loose. But always work at the foot of the cliff—never climb up them looking for samples.

GEOLOGIST *(left)*
Professional geologists sometimes work in places that would be far too dangerous for the amateur—but they always wear the appropriate safety equipment, including a hard hat.

hunting is to be considerate. Don't
ruin the site for others, clean out all
the samples, or damage the rock in an
attempt to extract your sample.

PLACER DEPOSITS

Some mineral crystals are tougher and
more durable than others and may
survive long after the rock in which
they were formed has disintegrated
through the ravages of time. If these
tough crystals are heavy too, they may
accumulate on shoals in streams and
rivers because they are the first to be
dropped when the current slackens.
Such deposits are known as placer
deposits. The best-known placer is
gold, but diamonds, garnets, and
emeralds can all be found like this.

You can search placer deposits for these valuable crystals by panning.
All you need for this is a broad, shallow pan—such as an old, lightweight
frying-pan with shallow sides. Scoop shingle from the bed of a likely looking
river, put it in the pan, and throw out any large stones. Then gently swirl a
small amount of water and the sand together in the pan, allowing a little
water and sand to swill over at each turn. With a little care, the lighter sand
and mud should be swilled out, leaving the heavier minerals behind. With a
great deal of luck, these may include a gold nugget.

MINERAL VEIN *(above)*
*By far the majority of the
rarest crystals of metal and
precious gems are found in
narrow bands in the rock
called "veins"—formed
in cracks in the rock
where unusual minerals
were concentrated.*

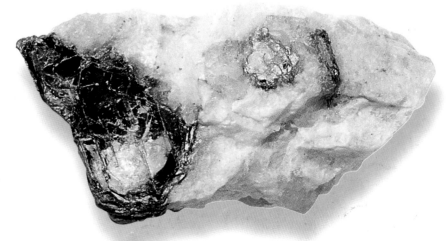

PEGMATITE *(left)*
*Narrow seams of rock
called pegmatites are
often sources of good
crystals of rare or
precious minerals—
because they form as the
last portion of magma
cools down (see page 15).*

Looking after your collection

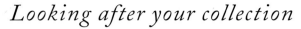

SINCE THEY COME out of the ground, most rock and crystal samples are fairly dirty when you find them, and they need to be cleaned to look their best in a collection.

For most minerals, the best way to clean them is to brush off loose dirt with a soft toothbrush, then wash them with warm (not hot) water. To shift greasy marks and stains, try adding a little detergent to the water. If the sample is coated with a hard layer of mud and grit, don't try to chip or scrape it off. Soak it in water overnight to soften it up so that it almost floats off. You can attack hard samples such as quartz with a nailbrush to get rid of stubborn dirt, but others such as gypsum and calcite are so soft or brittle that they can easily be scratched or broken.

Before washing, try and identify the mineral. Some, such as halite, are soluble and will be ruined by contact with water. For minerals like these, use a toothbrush. If the sample is very soft, just use a blower brush like those used by photographers.

Iron stains on certain minerals can be removed by soaking in oxalic acid. This is available from chemists but it is poisonous, so should be handled carefully. A few minerals are dissolved by oxalic acid, so try it out on a small fragment of the sample first. Vinegar can be used to dissolve calcite and other limey deposits on most insoluble minerals.

dust blower

awl

tweezers

CLEANING YOUR SAMPLES *(below) For proper presentation and display, rock and crystal samples need cleaning. Here are some of the items that you can use to clean your samples without damaging them.*

dusting brush

toothbrush

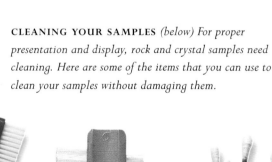

nail brush

cotton buds

LABELLING AND DISPLAYING

You may not have enough room to display more than a few prize samples, but it is worth carefully labelling, cataloging and storing your collection. You can get proper sample display drawers, but failing that, simply use shallow drawers, or a shallow cardboard or wooden trays in a cupboard. Identify every sample with a number written on either a sticky label or on a dab of white paint on an inconspicuous part of the sample. Use this number to enter the details of the sample in a catalog—either a computer file, a cardfile, or a notebook. Write the name and characteristics of the sample, and the date and place you found it.

Rather than just putting the samples in a drawer at random, try and arrange them in groups. You could group them by location or by their color, but most geologists prefer to group them by type. Rocks can be divided into igneous, sedimentary, and metamorphic, for instance, then fine-grained and coarse-grained. Mineral crystals can be grouped into chemical types, such as silicates or oxides, and silicates can then be divided into families such as the quartz family or the mica family.

DISPLAYING (above)
Make the most of your rock and crystal collection by displaying your samples properly in drawers or boxes with compartments.

LABELLING (left) Each sample should be identified with a mark or label, keyed into a card or computer file or logbook noting its identity and where it was found.

Identifying Quick Key

AT A GLANCE, all rocks look much the same—rather dull-colored, hard and lumpy. But it is surprisingly easy to make an educated guess at the type, once you know what to look for. The first clue with any rock is where you found it. The second step is to decide whether it is igneous, sedimentary or metamorphic.

IGNEOUS ROCKS
Look for tightly packed interlocking crystals—often slightly shiny. There should not be any bedding lines (sedimentary) or banding (metamorphic).

SEDIMENTARY ROCKS
Look for similar-looking grains, held together by a cement. They may even crumble as you rub them. Also look for bedding planes and fossils.

METAMORPHIC ROCKS
Look for wavy bands (foliation) or very smooth almost shiny dark or light surfaces.

Sedimentary rocks

THE FIRST STEP to identifying a sedimentary rock is to decide whether it is clastic (made from broken rock fragments) or chemical/biogenic (made from chemicals or the remains of living things). Chemical/biogenic rocks are usually whitish, creamy brown or light-grey in color. Clastic rocks tend to be brownish or grey-brown, dark grey or black.

If it is clastic, decide whether it is coarse, medium or fine-grained.

ARKOSE *Medium-grained*
MEDIUM-GRAINED clastic rocks may be sandstones, arkoses or graywackes.

SHALE *Fine-grained*
FINE-GRAINED clastic rocks include silts, shales, mudstones, and clays.

CONGLOMERATE *Coarse-grained*
COARSE-GRAINED clastic rocks may be conglomerates or breccias.

ACID

FINE

COARSE

RHYOLITE

QUARTZ PORPHYRY

GRANITE

TRACHYTE

ANDESITE

BASALT

Igneous Rocks

THE CHARACTER and identity of rocks depends on two main factors:

• How acid or basic was the magma it formed from. The more acid it is—that is the richer it is in silica—the lighter in color it is.

The igneous rocks in this panel are arranged so that the darkest, most basic rocks such as basalt and gabbro are on the right; the lightest, most acid rocks, such as rhyolite and granite are on the left.

DOLERITE

• How quickly it cooled and solidified, which means, essentially, how deep in the ground it formed. The deeper underground it formed, the slower it cools and the bigger the grains grow. The igneous rocks in this panel are arranged so that the finest-grained volcanic rocks, such as rhyolite and andesite are in the top row; medium-grained hypabyssal rocks such as quartz and dolerite are in the middle row; and coarse-grained plutonic rocks, such as granite, syenite and gabbro are in the bottom row.

Find the rock which matches your sample best.

SYENITE

DIORITE

GABBRO

Identifying minerals

MINERALS ARE MUCH HARDER to identify than rocks. One reason is that you rarely see separate crystals of a particular mineral. More often than not, they are mixed in with crystals of other minerals inside a lump of rock. Even when you do find large, separate crystals of a particular mineral, it can take many forms. To identify a mineral, you need to gradually narrow down the choice by going through the following tests.

WHERE DID YOU FIND IT?

To increase your chances of a quick identification, try to figure out what type of mineral it is from where you found it. The list below gives you an idea of the minerals you might find in particular places.

IGNEOUS INTRUSIONS
- quartz
- feldspar
- mica
- dark minerals

PEGMATITES AND CAVITIES IN LAVA
- quartz, feldspar and mica
- topaz
- beryl
- apatite
- tourmaline
- garnet
- sulphides

VEINS
- sulphides
- malachite
- azurite
- copper

VOLCANIC VENTS
- sulphur
- sulphates
- hematite

HOT SPRINGS
- travertine
- gypsum
- selenite

VOLCANIC DEBRIS
- pumice
- olivine
- augite

LIMESTONE QUARRY
- calcite
- dolomite
- gypsum
- fluorite
- galena
- sphalerite
- marcasite
- hematite

RIVER SANDS
- quartz
- gold
- cassiterite
- magnetite

METAMORPHIC ROCKS
- sulphides
- garnet
- mica
- calcite
- quartz
- spinel

Hardness

MINERAL HARDNESS is rated by the Mohs' scale against these standard minerals below. Find which standard mineral your sample will scratch, and which it can be scratched by, and rate it accordingly. If you do not have some of these minerals, use the substitutes suggested.

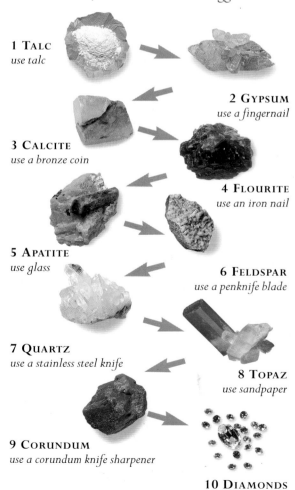

1 TALC
use talc

2 GYPSUM
use a fingernail

3 CALCITE
use a bronze coin

4 FLOURITE
use an iron nail

5 APATITE
use glass

6 FELDSPAR
use a penknife blade

7 QUARTZ
use a stainless steel knife

8 TOPAZ
use sandpaper

9 CORUNDUM
use a corundum knife sharpener

10 DIAMONDS

COLOR

Color gives an instant clue to a mineral's identity. The bright blue of azurite, for instance, is unmistakable. But traces of chemicals can turn many minerals all kinds of different colors.

AZURITE (below)

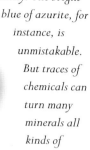

TRANSPARENCY

Some crystals are transparent, and you can see through them clearly like glass—like calcite, if it is pure. Others are translucent or milky, like fluorite, which means you can't see through them, but light shines through. Opaque crystals, like galena, block out the light altogether.

CALCITE (above)

STREAK

No matter what impurities it contains, a mineral will usually powder to the same color. Try using it to make a colored mark, called a streak, on the white, unglazed back of a porcelain tile and you may be able to identify it this way.

HEMATITE (above)

CLEAVAGE

Most minerals break apart or "cleave" more easily in one plane (direction) than others. The way it breaks is called the cleavage, and many minerals can be identified by their cleavage. Cleavage is often one of four types:

- one plane (like flat sheets), eg mica
- two planes (like a squarish rod), eg orthoclase feldspar
- three planes (like a block), eg halite
- four planes (like a diamond shape), eg fluorite

ORTHOCLASE FELDSPAR (above)

LUSTRE

Lustre is the way the mineral reflects light. The terms are vague, but lustre can be vitreous (shiny like glass), pearly, waxy, greasy, silky or adamantine (sparkly like diamonds).

Some, like opal, may also reflect shimmering rainbow colors. This is called iridescence.

OPAL & HALITE (above and right)

FRACTURE

Not all minerals have clear cleavage planes, but those like flint may break in these distinctive patterns:

- conchoidal (shell-like)
- hackly (jagged)
- splintery

FLINT (above)

ROCKS

Igneous volcanic rocks

EVERY YEAR HUGE amounts of red-hot, molten lava bubble from erupting volcanoes. As the lava flow spreads out from the volcano, it gradually cools and solidifies. Crystals grow within the cooling melt, and eventually it turns to solid rock. This lava-formed rock is a form of igneous rock called volcanic rock, or extrusive igneous rock. Lava is not the only source of volcanic rock. Any material spewing from an erupting volcano — ash, blobs of molten rock, and froth — can form volcanic rock if it turns to stone.

OBSIDIAN

Obsidian is one of the most beautiful of all rocks, prized long ago by the Aztecs and the Mayans. Shiny, translucent and jet black, it is actually volcanic glass. Occasionally, you may find it streaked with colors, like natural marbles — flecks of iridescent yellow, gold, and silver shimmering in the black, creating "gold sheen" or "rainbow obsidian" and various other varieties. Obsidian breaks like glass too, in conchoidal (shell-like) fractures with such sharp edges that the stone was highly valued by Native Americans long ago for arrowheads, and by the Aztecs for sacrificial knives. Obsidian forms when viscous (sticky), usually rhyolitic lava cools so quickly that there is no time for crystals to form. Obsidian tends to be found only where there has been relatively recent volcanic activity, because in time obsidian goes sugary and breaks down as it absorbs moisture. Old obsidian tends to dull as water gradually soaks in over thousands of years. Even in recently formed obsidian, only fairly fresh breaks look shiny. Sometimes cracks develop right through the rock, leaving the glassy cores surrounded by a bed of time-altered obsidian, called perlite. If the rock is worn away, it may leave the cores behind as smoky glass pebbles called "Apache tears."

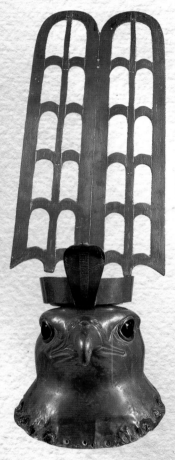

EGYPTIAN HEADPIECE (above) *Obsidian was treasured in Ancient Egypt and by the Ancient Aztecs for its dark beauty. It makes striking eyes in this gold falcon headpiece from the statue of the god Horus, found on the Ancient Egyptian site of Hierankonpolis.*

SNOWFLAKE OBSIDIAN (above, bottom) and **OBSIDIAN** (above center) *tends to form only in small outcrops. But there are some spectacular large outcrops, such as Glass Buttes in Oregon.*

RHYOLITE

*(above and left) One of
the most common of all volcanic
rocks, rhyolite forms from a very thick, acidic
magma that produces very explosive volcanoes.
Like most volcanic rocks, it is very fine-grained.*

ID CHECK:
VOLCANIC ROCKS

- **OBSIDIAN / PITCHSTONE**

TYPE: Volcanic rock

GRAIN SIZE: Glassy

GRAIN SHAPE: Vitreous, microcrystalline

COLOR: Black, brown

CHARACTERISTICS: Volcanic glass with
the same composition as rhyolite

- **RHYOLITE**

TYPE: Acid volcanic rock

GRAIN SIZE: Fine

GRAIN SHAPE: Granular, porphyritic

COLOR: Light gray, pink

CHARACTERISTICS: The fine-grained
equivalent of granite, solidifying
from acid, thick lava

PITCHSTONE

Pitchstone is indeed black as pitch (tar). It is similar to obsidian, but forms underground, where it is turned dark and duller by moisture. Indeed, some geologists distinguish between obsidian and pitchstone on the basis of their water content. Pitchstone has ten percent water, obsidian less than one percent. It gets its dullness from crystals of quartz and feldspar. Over time, all the glass will slowly turn to crystal, and the pitchstone will be indistinguishable from rhyolite.

RHYOLITE

Rhyolite is one of the most widespread of all volcanic rocks. It forms from the silica-rich (73 percent), highly acid magma typical of subduction zones — the same magma that forms granite when it solidifies underground. Viscous magmas like these tend to clog up volcanic vents, then burst through in a mighty, explosive eruption, pouring out streams of lava and huge clouds of ash, which is why rhyolites are usually found near tuffs, pumices and obsidians. The lava flows pile up thickly around the vent — too sticky to flow very far away — and it is from these flows that the rock forms. There are ancient rhyolite flows 800 feet (250m) thick in Nevada. Rhyolite is a fine-grained rock, typically light tan to pink in color, and the flow lines of the lava are often preserved in it, giving it a slightly banded appearance. Unlike granite, it often has bits of glass embedded in it.

PITCHSTONE *(right) is one of
the most easily recognized of all
volcanic rocks — easily
distinguishable by its dull,
pitch black color. It is similar
to obsidian, but contains
more water.*

**THE GREAT RIFT
VALLEY, KENYA** *(above)
The volcanic rock trachyte
often forms from the lava
that oozes up where
continents split apart — as
is starting to happen in
East Africa's Great Rift
Valley.*

TRACHYTE *(above) is
usually a very light-gray
color, typically with a
brown or reddish tinge —
perhaps dotted with dark
chips of mica.*

DACITE

Named after a Roman province in
Romania, dacite is a fine-grained
volcanic rock which can be
beautiful when polished, but is
often just used for road
chippings. It forms from fairly
sticky magmas (64 percent
silica). These are the lavas that
tend to form lava domes like
Mount Saint Helens in Washington
State. The main constituents are
pinky quartz and plagioclase feldspar,
though it has traces of biotite mica or
hornblende and pyroxene.

TRACHYTE

Trachyte is a volcanic rock that typically
forms from lava flows but is also found in dikes and sills. It is not as acid as
rhyolite nor as basic as basalt; it has a silica content of around 57 percent.
This is the kind of lava that oozes onto the surface as the vast plates that
make up continents are dragged apart, as along Africa's Great Rift Valley.
Lavas like these also gush onto the surface with basalt in island volcanoes
such as those of Iceland and Madagascar. A lava flow often leaves distinct
bands in the rock, known as the "trachytic texture."
Trachyte is a light-gray color, slightly darker than rhyolite.
Crystals of feldspar are set in a matrix of finer crystals. There are
also spots of dark minerals such as pyroxene and aegirine.

ANDESITE

The most common of volcanic rocks after basalt, andesites get their
name from the Andes mountains of South America, where they are
found in abundance. Andesitic lava is not as acidic as rhyolite, but it still
has a high silica content, so it forms a fairly sticky lava that tends to clog up
volcanoes before bursting through in an explosive eruption. Japan's Mount
Fuji is just one of the many largely andesitic volcanoes forming the Pacific
"Ring of Fire" — the ring of explosive volcanoes on the Pacific rim.
Andesitic magma was once part of the ocean crust before it was dragged
down into the Earth's interior by subduction, melted, and then burned its
way up through the crust again. It probably gets its silica content from sandy
seabed sediments that were pulled down with the ocean crust as it was

ANDESITE *(left) is usually much paler than basalt, and though almost as fine grained, it often contains large grains (called phenocrysts) of dark minerals which give it a salt-and-pepper look.*

subducted. Andesites are dark, fine-grained volcanic rocks, a little lighter in color than basalts, with a speckled salt-and-pepper look. They are rich in plagioclase feldspar, with smaller amounts of pyroxene, amphibole, and biotite mica.

BASALT

Basalts are the most common of all volcanic rocks. Indeed, 80 percent of all volcanic rock is basalt. They form from basic, very runny lavas that spread far out around a volcano. They often pour out of volcanic fissures in huge floods, forming plateaus thousands of yards thick, such as India's Deccan and the Columbia Plateau of Oregon. Because the lava cools so quickly, basalt rocks are very fine grained – so fine grained that the grain can sometimes only be seen under a microscope. They are typically very dark in color – even darker than andesite. Basalt is one of the most resistant to weathering of all stones, and is widely used as aggregrates for road-building, covered in tar. If you cannot see basalt in the landscape, the chances are you will find it, literally, in the street outside your home!

BASALT *(right) is a very fine-grained, dark (almost black) volcanic rock — and one of the most common of all rocks, found all over the world.*

ID CHECK:
VOLCANIC ROCKS

• DACITE
TYPE: Acid volcanic rock
GRAIN SIZE: Fine
GRAIN SHAPE: Granular with phenocrysts
COLOR: Medium gray, darker than rhyolite
CHARACTERISTICS: Volcanic glass with the same composition as rhyolite

• TRACHYTE
TYPE: Intermediate volcanic rock
GRAIN SIZE: Fine
GRAIN SHAPE: Porphyritic
COLOR: Medium gray, pink, slightly darker than rhyolite
CHARACTERISTICS: Light-colored volcanic rock rich in feldspar

• ANDESITE
TYPE: Intermediate volcanic rock
GRAIN SIZE: Fine
GRAIN SHAPE: Porphyritic
COLOR: Dark, gray-black
CHARACTERISTICS: More basic than trachyte and darker in color

• BASALT
TYPE: Basic volcanic rock
GRAIN SIZE: Fine
GRAIN SHAPE: Dense, porphyritic
COLOR: Dark, gray-black
CHARACTERISTICS: Dark, fine-grained and very common

TUFF

Tuffs are essentially volcanic ash, cemented together over time by lime or silica washed down through the rock. Looking a bit like brown bread, they are usually riddled with holes – which is why they are often used for insulation – and the grain size is wildly variable, with powdery grains often mixed in with giant lumps. Often, though, the grains are layered, because the larger, heavier grains fell to the ground first. You may find augite and feldspar crystals embedded within tuff.

IGNIMBRITE

When a volcano erupts, pressure in the lava is suddenly released and gas bubbles froth up explosively through the melt. Sometimes, these bursting gas bubbles may drive a glowing, red-hot cloud of molten lava drops and ash roaring down the mountainside faster than a jetplane. This deadly, high-speed cloud, called a *nuée ardente*, incinerates everything in its path. When such an avalanche swept down on the city of Saint Pierre in Martinique, when nearby Mont Pelée erupted in 1902, the entire city was scorched out of existence in minutes. The tremendous heat of these avalanches eventually welds the lava drops and ash together into a solid rock called an ignimbrite.

TUFF *(above), also called tephra, is a word that comes from the Latin for "porous stone," and it is indeed full of holes created by the gas bubbles trapped in the volcanic ash as it solidified. Tuffs vary according to the nature of the volcanic lava from which they formed. Andesite tuff forms from andesite lava, rhyolite tuff from rhyolite lava. Solid lumps of tuff vary from nut-sized "lapilli" to brick-sized "blocks" and "bombs."*

IGNIMBRITE *(above) is a word that comes from the Latin for "fire cloud," and this rock forms from the glowing clouds bursting from volcanoes.*

PUMICE

Pumice is the only rock which floats. It is a light-gray rock made from the froth that often forms on the top of lava and is so completely riddled with holes made by gas bubbles that it is foam light. Froth occurs most readily when the lava is sticky, so pumices are typically linked with rhyolites – most famously on the island of Lipari in Italy. When the giant volcano Krakatoa in Indonesia erupted in 1883, it threw out huge blocks of pumice which floated right across the Indian Ocean and were a hazard to shipping for months afterwards. Some people use pumice as an abrasive for removing dead skin, and dentists sometimes use powdered pumice to scour teeth.

ID CHECK:
VOLCANIC ROCKS

• TUFF
TYPE: Volcanic rock
GRAIN SIZE: Fine—coarse
GRAIN SHAPE: Granular fragmental, glassy
COLOR: Gray, buff, mottled
CHARACTERISTICS: Finer grained, volcanic fragments made from welded ashes

• IGNIMBRITE
TYPE: Volcanic rock
GRAIN SIZE: Fine—coarse
GRAIN SHAPE: Granular—glassy
COLOR: Speckled, white, buff
CHARACTERISTICS: Rocks forming from glowing avalanches

• PUMICE
TYPE: Volcanic rock
GRAIN SIZE: Fine
GRAIN SHAPE: Vesicular
COLOR: White-gray
CHARACTERISTICS: The only rock which floats

KRAKATOA *(left) When the volcano Krakatoa in Indonesia erupted in 1883, the eruption flung out huge amounts of pumice, which floated thousands of miles across the Indian Ocean.*

PUMICE STONE *(below) Pumice's rough texture makes it a good astringent, and people have used pumice stones to clean the skin for thousands of years.*

PUMICE *(right) forms from the froth of lava and looks like light foam. The holes formed by the gas bubbles make it so light it actually floats.*

Igneous hypabyssal rocks

SMALL POCKETS of magma that cool and solidify just beneath the surface, typically in dikes and sills, form what are often known as hypabyssal rocks. Because they are nearer the surface and in smaller pockets, they cool much more quickly than deeper masses of plutonic rock. Crystals do not have time to grow very big, so the rocks are generally finer grained – though not as fine grained as volcanic rocks which cool even faster on the surface. The distinction between hypabyssal and plutonic rocks, though, is not all that clear, and some geologists prefer to talk of fine-grained or "micro" versions of the plutonic.

PHAROAH *(below) Many of the fantastic statues of Ancient Egypt have retained their sharp features for thousands of years because they were carved from the tough igneous rock quartz porphyry.*

QUARTZ PORPHYRY

Quartz porphyry, or microgranite, is the pale, tough rock from which many Ancient Egyptian statues were carved. It typically forms in dikes and sills and is the hypabyssal equivalent of granite deep down and rhyolite on the surface. What makes quartz porphyry different is that amid the medium-size grains are extra-large crystals, or "phenocrysts," of quartz. Any igneous rock that contains phenocrysts is said to be "porphyritic." Various hypabyssal rocks besides quartz porphyry are porphyritic, as are some volcanic and plutonic rocks.

MICROGRANITE
(right) is also known as quartz porphyry. It is formed from the same magma as granite and rhyolite. but solidifies just below the surface.

PEGMATITE

Hot magma deep underground is packed with minerals which gradually form crystals as the magma cools and solidifies. Minerals with the highest melting points crystallize out first, so the chemical mix of the remaining molten magma gradually changes as it oozes upwards. In this way, the concentration of unusual minerals gradually increases as the magma seeps up through cracks in the rock. When the last vestiges of the melt finally crystallize, they form rocks called pegmatites.

Among the crystals that form in pegmatites may be rare gems such as tourmaline, aquamarine, chrysoberyl, topaz, emeralds, and garnets. Pegmatites are the main source of beryls. Some of the crystals found in pegmatites are genuinely gigantic – including single beryl crystals the size of telegraph poles and giant sheaves of mica as big as a car. Pegmatites are named after their plutonic equivalent (e.g. granite pegmatite) or the minerals they contain (mica pegmatite). Often when geologists talk about pegmatites, they mean granite pegmatite.

DIABASE

Diabase, also known as dolerite, is the medium-grained hypabyssal equivalent of basalt and gabbro – in other words, a very basic rock. Diabases are dark like basalt, but can be distinguished by their slightly larger grain, which gives them a slightly mottled look. They are typically found in dikes, where they form from basalt lava. Diabase sills can be huge, as in the Palisades of New Jersey, which are 1,000 yards (300m) thick in places and stretch for 50 miles (80km) or more along the Hudson River. Diabase is a tough stone used for roads. The Bluestones of England's ancient stone circle Stonehenge were carved from diabase dug from the Prescelly Mountains in Wales.

PEGMATITE *(above) is one of the least common of all igneous rocks and forms only from the last globs of molten magma to cool – but it often contains a rich range of rare and precious gems.*

ID CHECK:
HYPABYSSAL ROCKS

- **QUARTZ PORPHYRY**

  TYPE: Acid igneous rock

  GRAIN SIZE: Medium

  GRAIN SHAPE: Porphyritic with phenocrysts

  COLOR: Light, pink, speckled

  CHARACTERISTICS: Contains phenocrysts of quartz

- **GRANITE PEGMATITE**

  TYPE: Acid igneous rock

  GRAIN SIZE: Coarse–very coarse

  GRAIN SHAPE: Granular, coarsely porphyritic

  COLOR: White, pink, reddish

  CHARACTERISTICS: Typically occurs in veins, often with crystals growing parallel

- **DIABASE**

  TYPE: Basic igneous rock

  GRAIN SIZE: Medium

  GRAIN SHAPE: Equigranular, porphyritic

  COLOR: Dark, greenish-gray

  CHARACTERISTICS: The hypabyssal equivalent of basalt and gabbro

DIABASE *(right) is also known as dolerite, and is indeed a very basic – non-acidic – igneous rock. This is why it is dark like other basalts, but it sometimes has a greenish tinge. Most dolerites are very old.*

Igneous plutonic rocks

PLUTONIC ROCKS are igneous rocks that form from magma deep underground, typically in large dome-shaped masses called batholiths. It takes a long time for such giant masses of hot magma to cool, insulated from the air by thick layers of rock above. Crystals have plenty of time to grow, and so plutonic rocks are all coarse grained.

GRANITE

Granites are the most common of all the plutonic rocks, forming far underground as batholiths. Although they only form far beneath the ground, they are often seen on the surface because they are so tough that they survive long after the softer rock around them has been weathered away – leaving the granite standing out in massive outcrops, like Río de Janeiro's Sugarloaf Mountain, Yosemite's Half Dome, Georgia's Stone Mountain or the tors on England's Dartmoor. At Mount Rushmore, granite has been carved into the huge faces of the presidents in the cliff. Because it is so tough, it makes a very durable building material – and polished granite covers many an expensive new building.

Granite is a light-colored, speckled rock made from a mix of white or pink feldspar, quartz and small amounts of biotite mica or muscovite. The grains are large, with feldspar crystals sometimes growing 4 inches (10cm) long.

GRAPHIC GRANITE
(above) Granite is the most common of all the plutonic rocks – that is, igneous rocks that form as magma solidifies deep underground.

WHITE GRANITE *(left) Because it cools from magma slowly at depth, crystals have plenty of time to grow, so granite is usually very coarse grained.*

GRANITE *(left) is beautiful when sliced and polished, and big slabs of polished granite are often used to dress the walls of more expensive office buildings.*

SYENITE

Syenite gets its name from Syene in Egypt where it was quarried over 5,000 years ago for building stone. In the United States, it is found in New Hampshire and Massachusetts, among other places. It is the coarse-grained, plutonic equivalent of trachyte. Although syenite is much less common, it looks a little like a darker version of granite. The difference is that syenite contains orthoclase rather than plagioclase feldspar and much less quartz. There is an especially attractive variety of syenite called larvikite, which is found near Larvik in southern Norway, which shimmers blue-green when polished.

SYENITE (above) looks rather like granite, but contains little or no quartz.

DIORITE

Diorite is the coarse-grained, plutonic equivalent of andesite and forms in the same places. Huge masses of diorite are found in Alaska and the Rockies. It is much less common than granite and typically forms when granitic magma is contaminated by impurities. The result is that it is often seen in offshoots of large granite masses. It is basically a black and white rock, lighter in color than the similar gabbro, and is popular for cobbles. It consists principally of plagioclase feldspar and hornblende and can be distinguished from granite by its lack of quartz.

MOUNT RUSHMORE (below) *The famous giant heads of American presidents have been carved from the solid granite of Mount Rushmore.*

DIORITE (right) *is a black and white rock that typically forms when granite is contaminated by impurities from the surrounding rocks.*

ID CHECK:
IGNEOUS PLUTONIC ROCKS

• **GRANITE**

TYPE: Acid plutonic rock
GRAIN SIZE: Coarse
GRAIN SHAPE: Granular, porphyritic
COLOR: Light, gray, pink
CHARACTERISTICS: The coarse-grained equivalent of rhyolite

• **SYENITE**

TYPE: Intermediate plutonic rock
GRAIN SIZE: Coarse
GRAIN SHAPE: Equigranular, porphyritic
COLOR: Pinkish or white
CHARACTERISTICS: Looks very like granite, but slightly darker

• **DIORITE**

TYPE: Intermediate plutonic rock
GRAIN SIZE: Coarse
GRAIN SHAPE: Equigranular, porphyritic
COLOR: Speckled black, green
CHARACTERISTICS: The coarse-grained equivalent of andesite

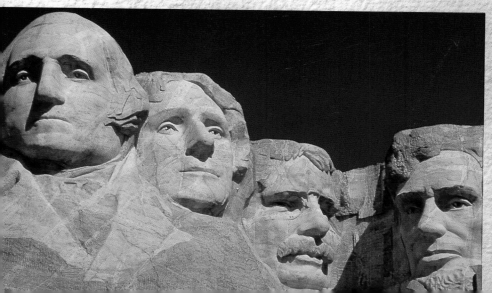

GABBRO

Taking its name from the Italian village of Gabbro, this rock is very similar to diorite but darker in color. Although usually black and white, it can contain green olivine crystals, and often takes on a stripy appearance as pyroxene and plagioclase form in layers within it. Gabbro typically occurs in big intrusions – stocks, bosses and dikes, and lopoliths, which can sometimes extend over huge areas, as in Montana.

GABBRO *(above) is a black and white plutonic rock very similar to diorite, named after a village in Tuscany in Italy, one place where the rock is found.*

PERIDOTITE

Peridotite may well be the most abundant rock in the world. Indeed, a large proportion of the world may actually be made of peridotite. The echoes of earthquake waves and various other clues indicate that the Earth's mantle – that is, all of its interior but for the thin crust and the small core – is probably made of peridotite. But it only rarely comes to the surface. When it does, we can see it is an ultrabasic rock, coarse grained and made mostly of olivine. It typically forms when olivine crystals precipitate from molten basalt, or in nodules brought to the surface in basalt lavas.

PERIDOTITE *(above) is probably the rock of which most of the Earth's mantle is made, but it is uncommon on the surface.*

GEM PERIDOT *(left) Both peridotite and the green gem peridot form mainly from the mineral olivine. The gems form in basalt lava and are quite rare.*

QUAKE ROCK *(right) If it were not for the massive movements of the Earth's crust that cause earthquakes, peridotite would probably not appear on the surface*

DUNITE

Dunite is a form of peridotite made almost entirely of olivine, which gives it a distinctive greeny-brown color. It gets its name from Mount Dun in New Zealand.

DUNITE *(left and above left) The high proportion of olivine in dunite makes it almost as green as olive oil. In fact, it can be almost pure olivine, as at Webster in North Carolina.*

Clastic sedimentary rocks

CLASTIC SEDIMENTARY rocks form from fragments of rock weathered and eroded away by rivers, glaciers, wind, and waves over the ages. Washed and blown along by rivers and wind, these fragments settle on the sea floor, on river and lake beds, on floodplains and beaches and in deserts. There they slowly pile up and become compacted into rock over millions of years.

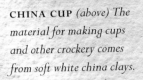

CHINA CUP *(above) The material for making cups and other crockery comes from soft white china clays.*

LUTITES: CLAYS, SILTSTONES, SHALES, AND MUDSTONES

Lutites are made up from powder-fine sediments – grains less than 0.002 inches (0.05mm) – typically of quartz and clay minerals.

In clay, the grains are so tiny that you can only see them with a magnifying glass. The result is that, when pure, clay feels smooth and slippery, like plasticine. It is

CLAY *(left) Clay is the finest-grained rock of all – so fine that when pure, it feels smooth and slippery. It tends to hold on to water, and when wet, it can be moulded easily.*

BANDED SILTSTONE *(above) varies considerably in color, depending on the minerals it contains, from black to light brown, and may often show distinct bands.*

SILTSTONE CLIFF *(left) Where it is exposed on the coast, siltstone forms dark, smooth, chunky blocks with surprisingly flat horizontal or tilted surfaces.*

SHALE *(right) is one of the most common sedimentary rocks. It is dark gray, very fine-grained and has a slate-like parallel structure but does not break like slate.*

very soft, and can easily be moulded when wet. The minerals in clays are not only rock fragments, but may be mixed in with the remains of seashells and plants. Clays come in many different colors, from white china clays used for making crockery, to gray clays rich in plant material, and rust-colored clays rich in iron.

In siltstone, many of the grains are just visible to the naked eye, and have a slightly gritty feel when rubbed. Whereas clay is mostly clay minerals, siltstone has a higher proportion of quartz and mica. It is a relatively soft rock and can sometimes crumble away in your fingers. You can often see ripple marks in it from water that washed over it as it formed on sea and river beds.

Shale is formed by layer upon layer of mud settling on the sea floor or in shallow lakes, which was squeezed into laminations, which are like the pages of an ancient book. Shale looks flaky like slate, but shale is much softer and often contains the fossilized remains of clams and snails, buried in the mud and preserved forever as the mud turned to stone. The color of shale is very variable and depends on the minerals it contains. Black shales are rich in carbon from plant remains. Oil shales are black shales rich in bitumen.

As their name implies, mudstones form from mud. They are similar to clay, but mudstones tend to have a much higher carbon content — the altered remains of plants and animal skeletons — and often contain fossils. You can test for the carbon in mudstone with very dilute hydrochloric acid.

SILTSTONE *(above right) The grains in siltstone are very fine, like those in clay — but are typically just big enough to be seen by the naked eye.*

ID CHECK:
CLASTIC SEDIMENTARY ROCKS

• **SILTSTONE**
TYPE: Clastic argillaceous
GRAIN SIZE: Fine
GRAIN TYPE: Laminar
COLOR: Gray, buff, with white specks
CHARACTERISTICS: Clay-like rock with grains from 0.0625 to 0.004 mm

• **SHALE**
TYPE: Clastic argillaceous
GRAIN SIZE: Fine
GRAIN TYPE: Laminated, flaky with lumps
COLOR: Buff or black with darker inserts
CHARACTERISTICS: Often rich in clam fossils

• **MUDSTONE**
TYPE: Clastic argillaceous
GRAIN SIZE: Very fine
GRAIN TYPE: Clay
COLOR: Variable
CHARACTERISTICS: Looks like dried mud

ARENITES: SANDSTONES, ORTHOQUARTZITES, ARKOSES, AND GRAYWACKES

Sandstone lives up to its name. It is literally stone made from grains of sand. Sometimes the sand was laid down by desert winds, and the grains were worn almost round as they were buffeted across the dry landscape. Sometimes the sand was laid down on river beds, beaches, and seabeds, in which case the grains are less worn and more angular. Sandstones formed from glacier debris, or high up in rivers where the sand has not travelled far, are sharpest of all.

They get their color and texture from the material that cements the grains together. Red and brown sandstones are stained by iron-rich hematite which gives the familiar color to New York's famous "brownstone fronts", and the rusty red monuments of the mesas and buttes of Utah, Colorado and Arizona. White sandstones are cemented by calcite, and it is these iron-free sandstones, like those near Crystal City, Missouri, that are used for making glass. Yellow sandstones get their hue from limonite. Blue and black sandstones, like those of Alberta, are colored by bitumen and carbon. Sometimes the cement is so weak that the stone crumbles in your fingers. Most of the time, though, it is much harder, and turns sandstone into perfect building material – the most widely used building stone of all.

Most sandstones are rich in history. Cracks and ripples in bedding planes bear witness to the long ages over which the sediments were laid down. You can see cracks where the sand dried out in the sun and wavy lines where ripples of water washed over it. You can see the grains of sand graded as the current slowed down. Desert sandstones reveal the shapes of dunes from millions of years ago, like those at Zion Canyon, Utah. You can also find fossils of small creatures that once lived in the sand or the waters above. There may even be bones or footprints of larger creatures, including dinosaurs – like the huge footprints found in the sandstone quarries at Portland, Connecticut.

Sandstones fall under the heading of arenites, which describes any sedimentary rock with medium-sized grains, between 0.002 inches (0.06mm) and 0.08 inches (2mm) across. These grains are clearly visible to the naked eye and give the rocks a rough feel. Orthoquartzite sandstone

SANDSTONE *(above) is a blanket term describing a wide range of rocks made largely from grains of sand. They can form anywhere from deserts to deep beneath the sea.*

ORTHOQUARTZITE *(above) formed from sediment deposited on the sea bed by rivers running into shallow seas.*

ERODED SANDSTONE *(above) Many sandstones formed long ago from the sand of the huge deserts of certain periods in the past, such as the Permian, 290 million years ago.*

is made largely of quartz grains deposited in shallow seas. Arkose is a pinkish sandstone rich in feldspar, which indicates that it must have formed in the desert, since feldspar rots in damp places. Graywackes get their name from the German word *Grauwacke*, which means "gray grit." They are dark gray-green sandstones made from sand embedded in clay and mica. Many were probably laid down by turbidity currents — violent undersea avalanches set off by storms and earthquakes that roar down canyons on the seabed. There are huge graywacke deposits in the Appalachians.

ID CHECK: ARENITES

• **SANDSTONES**

TYPE: *Clastic arenaceous*

GRAIN SIZE: *Fine–coarse*

GRAIN TYPE: *Granular*

COLOR: *Buff to rust and gray*

CHARACTERISTICS: *Look like dark, hardened sand, but may be tinged with rust red*

ARKOSE *(above) is a sandstone tinged pink by feldspar. Arkoses are rocks of predominantly dry regions.*

GRAYWACKE *(right) is a dark-gray rock that forms where undersea currents pile up mud on the sea floor.*

RUDITES: CONGLOMERATES AND BRECCIAS

When sediments are made from gravel and pebbles bigger than 0.08 inches (2mm) across, they are called rudites.

Conglomerates are rudites made from big, smoothly rounded pebbles. These pebbles survive because they are made from tough materials like quartz, flint, chert, and various igneous rocks. But wearing them round takes a lot of time and effort, and they were probably washed up and down beaches or along shingle beds in fast-flowing rivers for hundreds of thousands of years before being consolidated into conglomerates. Often the dark pebbles contrast with the light cement around them, making them look like sultana puddings and earning them the name "puddingstone." The puddingstones of Roxbury, Massachusetts are well known.

Some of the biggest conglomerate deposits are left by rivers fanning out into the plains in the desert, like those in Death Valley, California. In Van Horn, West Texas, there are conglomerate deposits hundreds of yards thick.

Breccias are made from equally big pebbles as conglomerate, but the pebbles are angular, not smooth and round. Breccia can contain almost any kind of rock, but is nearly always found fairly near its source, since the pebbles would be sorted by water if washed any further away. In mountain areas they probably form from screes (rock falls) which gradually get cemented together by fine material accumulating between the stones. In deserts, they are probably created by flash floods.

A few breccias are formed volcanically, as volcanoes chuck out large fragments of volcanic rock. Others are formed by earthquakes, or land-slides, or along fault lines, where rock is crushed by the movement of the ground. Some breccias form underground as buried beds of rock are crushed by the weight of overlying rock.

CONGLOMERATE (*above*) *forms largely from beach material – and is immediately recognizable by the huge, smoothly rounded pebbles it contains.*

PUDDINGSTONE (*right*) *Dotted with round pebbles, conglomerate looks almost like a currant pudding, which is why it is sometimes called puddingstone.*

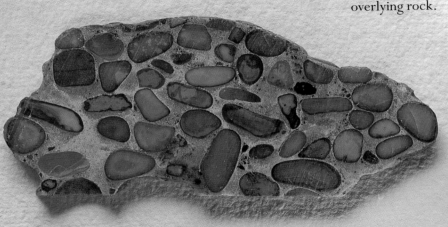

VOLCANIC CHUNKS *(above) Breccia can form in a number of different ways — including from the chunks of lava flung out by an erupting volcano.*

BRECCIA *(right) is instantly identifiable from the big, sharp, rock fragments it contains — distinct from the rounded pebbles of conglomerate.*

Biogenic, bioclastic, and chemical sedimentary rocks

CLASTIC SEDIMENTARY rocks are formed from tiny fragments of broken rock, and you can often see traces of the original rocks and minerals in the grains. But chemical and organic sedimentary rocks have a powdery texture that bears no traces of rock fragments. Chemical sediments are made from dissolved minerals settling out of water. Organic sediments are made by living things and their remains. Most sediments contain at least some fossils, but organic sediments are made entirely of the remains of living things. They are of two types: bioclastic and biogenic. Bioclastic rocks are made from the fragmented remains of plants and sea creatures. Biogenic rocks are actually made by living creatures, like corals.

LIMESTONE *(above) is the most common of all organic sedimentary rocks — formed almost entirely from the altered remains of living organisms.*

LIMESTONE

Made almost entirely from one whitish mineral, calcium carbonate or calcite, limestones are very distinctive. Christopher Wren used white Portland limestone to build many of London's churches, including Saint Paul's Cathedral, after the Great Fire of 1666. Thousands of years earlier, limestone was used to build the great pyramids of Ancient Egypt.

Limestones form almost entirely in shallow, clear tropical

THE PYRAMIDS *(above), built by the pharaohs of Ancient Egypt over 3,000 years ago, are constructed from large blocks of limestone, cut and hauled into place by thousands of slaves.*

LIMESTONE *(below left) comes in many forms but most are a dull gray-white — the color of the mineral calcium carbonate.*

waters — but this does not mean they are only found in the tropics, because the continents have shifted so much through the ages that places once in the tropics are now in the Arctic Circle.

CORAL ROCK *(right)*
In the past, coral reefs may have been more extensive than they are today and reef limestones form from the remains of creatures that lived on coral reefs.

In some limestones, the calcite came mainly from sea shells, coral, and algae once living in warm tropical waters. In others, the calcite settled out of sea water chemically. Most contain a mix of organic and chemical calcite. But a few, like shelly limestone and chalk, are made almost entirely of fossils. In Bermuda, many people build their houses from coquina, the local limestone, which is essentially bits of shell cemented together, and you can often see recognizable shells of creatures like sea lilies, brachiopods, and corals within it.

Reef limestones are made entirely from the remains of creatures that live on coral reefs, including the corals themselves. Often they occur in layers of shelly limestone — a reminder that the reef was once set within wider seas. Because they are harder than shelly limestones, they may be left protruding as small hills called reef knolls as the shelly limestone is weathered away.

Limestone is often broken into large blocks like giant bricks by cracks called joints. All rainwater is slightly acid, and limestone is very susceptible to corrosion by acid. So rainwater seeping into limestone joints etches out potholes, deep gorges and even vast caverns underground — all typical features of "karst" country, named after the Karst limestone region of former Yugoslavia.

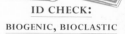

ID CHECK:
BIOGENIC, BIOCLASTIC

• **LIMESTONE**
TYPE: Organic carbonate
GRAIN SIZE: Fine, variable
GRAIN TYPE: Granular, often with fossils
COLOR: Gray, white, bluish
CHARACTERISTICS: Usually rich in fossils including shells

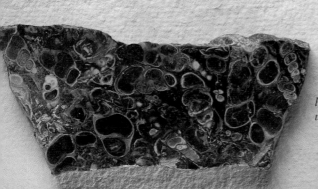

SHELLY LIMESTONE *(left and right) Limestones are made mainly from the remains of sea creatures and their shells — and often contain recognizable fossils of the shells, preserved within the rock.*

CHALK

Chalk is a soft, white rock of almost pure calcite found in Europe and North America. In the Cretaceous period – some 100 million years ago, when dinosaurs roamed the Earth – billions of tiny algae-like plants called coccoliths grew across the seabed about 200 fathoms down, undisturbed by storms above. These tiny plants, along with almost equally small shells of tiny creatures called foraminifera, gradually turned to tiny discs of calcite when they died, and layer upon layer of these microorganisms, along with the occasional larger shells, built up into thick, white chalk star – the white rock that forms the famous White Cliffs of Dover, and the Texas hills.

CHALKS

(above and right) Still often used in schools for writing on blackboards, chalk was once made from chalk rock – a soft white form of almost pure calcite made from the fossilized remains of microscopically small sea creatures called coccoliths.

FLINT AND CHERT

Sometimes in limestone, and especially chalk, you may see little hard balls or nodules of flint and chert. Bubbles of silica-rich fluids often form in volcanic material or in sediments from the remains of sea sponges. Flints and cherts form when these bubbles turn solid. They look like white pebbles on the outside, but once broken they look like dark toffee-colored glass, although they are incredibly hard. They break with conchoidal (shell-shaped) fractures, with a glass-like edge so sharp that they were used as axes, arrowheads, and knives in the Stone Age. Flints were also used to strike sparks to ignite the gunpowder in early guns.

FLINT

(above) Made from bubbles of silica-rich fluids which turn very hard and glassy – which is why they were widely used in the Stone Age for making tools such as axeheads.

CHERT

(right) Silica-rich fluids can harden into other forms besides flint nodules, though geologists are not entirely sure how they form. These rocks are called cherts.

OOLITIC LIMESTONE

Oolitic limestones are fossil-rich rocks made from tiny balls of calcite called ooliths, made as grains of silt rolled back and forth in lime mud. The word oolith comes from the Greek *oon*, meaning "egg," and, sometimes, a sheet of ooliths can look like fish roe, earning the rock the name roe limestone. There are oolitic shoals where ooliths are forming even today beneath the sea off the Florida coast near the Bahamas.

DOLOMITE

Dolomite or dolostone looks very much like limestone, and it probably started off in much the same way. But it is actually chemically different, and generally softer, which is why the dolomite facing of London's Houses of Parliament has had to be renewed as it is attacked by the city's polluted air. While limestone is mainly calcium carbonate or calcite, dolomite stone is mainly the mineral dolomite or calcium magnesium carbonate. In other words, some of the original calcium has been replaced by magnesium, a process called dolomitization. This probably happened as seawater seeped into the original limestone. Sometimes, spots of hard, unchanged limestone are left within the soft, new dolomite. You can test the difference with dilute hydrochloric acid. Limestone will fizz and bubble in acid; dolomite will do so only when powdered.

OOLITIC LIMESTONE *(above) can look almost like fish-roe, because it is made from tiny balls of calcite — formed as grains of silt were rolled back and forth in lime-rich mud.*

ID CHECK:
BIOGENIC, BIOCLASTIC

• **CHALK**
TYPE: Organic carbonate
GRAIN SIZE: Very fine
GRAIN TYPE: Chalky, powdery
COLOR: White, yellowish
CHARACTERISTICS: Powdery white rock containing microscopic marine fossils

• **OOLITIC LIMESTONE**
TYPE: Organic carbonate
GRAIN SIZE: Fine–medium
GRAIN TYPE: Rounded like fish roe
COLOR: White, yellowish
CHARACTERISTICS: Distinguished by its fish-roe grains

• **DOLOMITE**
TYPE: Chemical carbonate
GRAIN SIZE: Fine–coarse
GRAIN TYPE: Crystalline, sugary
COLOR: Yellow, gray
CHARACTERISTICS: Shelly limestone altered chemically as magnesium replaces calcium

DOLOMITE *(right) Unlike other limestones, dolomite is mainly magnesium carbonate, not calcium carbonate.*

TUFA, TRAVERTINE, AND EVAPORITES

Not all sedimentary rocks form from debris settling in water or blown by the desert. Some form when mineral rich waters evaporate, leaving behind solid deposits.

Tufas are deposits left around the rim of cool, calcite-rich water, rather like the limescale in baths and taps in areas of hard water. In places, these calcite-rich waters may drip from the ceilings of caverns, leaving long, hanging, icicle-like stalactites, or drip on the floor, building up tall pillars called stalagmites. These "dripstones" can build up in all kinds of fantastic formations besides stalactites and stalagmites. Sliced across, they reveal how they were built up in layers, like the layers of an onion.

Around hot springs, a denser, harder crust often builds up as the water evaporates, making the rock travertine. It is a pale honey color, with delicate banding, and often used by sculptors as a substitute for marble. It is also cut into slabs and made into polished floors. The most famous travertine is Roman Travertine, which gave the rock its name.

Wherever salty water evaporates, it can leave behind its dissolved minerals as deposits called evaporites. This can happen when salt lakes dry up, as in the Great Salt Lake, Utah. But it usually happens on a much larger scale when sea water evaporates in shallow seas and lagoons. In the Miocene epoch between five and 23 million years ago, the entire Mediterranean Sea became a site of evaporite formation as the Straits of Gibraltar were blocked off.

Evaporites are soft and easily worn away. So the best preserved evaporites are often in deserts where there is little water to wash them away. The desert rose is a weird and wonderful evaporite formation made of gypsum, the mineral used for plaster

TUFA (top) forms in the same way that limescale forms in kettles and in bathtubs in areas of hard water — as calcite precipitates from the water.

TRAVERTINE (above) is harder and denser than tufa and is usually honey colored. It forms from hot springs rather than cold water.

EVAPORITES (above and left) form when salty water evaporates, leaving behind a deposit of the minerals that were dissolved within it.

of Paris. The other main mineral in evaporites is rock salt. But the type of sediment created depends on the temperatures involved and the concentration of mineral salts in the water.

STALACTITES (above) are a rock called dripstone which forms as calcite-rich waters drip from cavern roofs in limestone areas.

COAL

Coal and peat are rocks that burn because they are made of fossil wood and plant remains. Most of the world's coal was formed 350 million years or so ago in the Carboniferous Period. At this time, large areas of the world were covered in dense, tropical swamps where giant club mosses and tree ferns grew in profusion. Over millions of years, layer upon layer of dead plants sank into the mud and did not rot due to the absence of oxygen. Instead, they were gradually squeezed and dried by the weight of plant remains and mud above. As they were packed down further and further, they gradually turned to increasingly pure carbon.

At the top is soft, brown peat (60 percent carbon) in which plant remains can still be identified and which burns with a very smoky flame. Below that is harder, brown coal or lignite (73 percent carbon), in which plants can only be identified in places. Below that is black, bituminous coal (83 percent carbon) in which plants can almost never be identified. At the very bottom is high-grade coal called anthracite – highly compressed, hard, and almost pure black carbon (94 percent) – which burns with an almost smokeless flame. It is thought that about 6 inches (15cm) of peat are squeezed to form 0.4 inches (1cm) of anthracite.

BROWN COAL *(top right),* **ANTHRACITE** *(far right) and* **BITUMINOUS COAL** *(right) form from the compressed remains of swamp trees over hundreds of millions of years.*

Metamorphic rocks: contact

ROCK COMING in direct contact with red-hot magma may be seared beyond all recognition. The sheer heat may re-form and realign crystals in the rock so dramatically that it becomes, in effect, a new rock, just like cooking changes damp cake mixture into fluffy cake. This is contact metamorphism. The crown of seared rock around a batholith is called the aureole. The closer they are to the magma and the bigger the intrusion, the more the rocks are altered. Just what rocks they are altered to depends on the original rock.

HORNFELS *(above) is a dark rock that forms when shale and mudstone are transformed by the heat of an igneous intrusion.*

HORNFELS AND SPOTTED ROCK

Metamorphism affects clays and shales in different ways. Right next to the magma, it is so hot that shale is completely recrystallized, forming a very fine-grained, dark, hard, splintery rock called hornfels. Sometimes garnets may be embedded in the rock. Further away, it is not quite so hot, and not all the minerals are recrystallized, leaving the rock spotted with unchanged minerals and so called "spotted" rock. Nearer the heat of magma, the spots are often crystals of the silicate mineral andalusite; further away they are more likely to contain mica crystals as well.

CARRARA MARBLE *(left) The Carrara quarry in Tuscany, Italy has been famed for its brilliant white marble since Roman times and was a favorite source of stone for the sculptor Michelangelo, who used it for his statue of David.*

SPOTTED ROCK *(right) is a kind of hornfels that forms a little way away from an igneous intrusion—where some minerals are left unchanged, forming spots within the rock.*

MARBLE HEAD (right)
*The pure color and
beautiful, smooth texture
of marble make it a favorite
stone with sculptors.*

MARBLE

When limestone is cooked by contact
metamorphism, the cake that emerges from
the oven is the beautiful rock, marble, as the
calcite is recrystallized into an almost sugary
texture. Pure limestone gives brilliant
white marbles that look like shiny
icing sugar. The most perfect is
the Carrara marble from the Italian Appennine Mountains,
cherished by sculptors such as Michelangelo and Bernini
for carving statues from. Impurities in the original
limestone may give all kinds of colored streaks
within the marble which can look beautiful
when polished as table tops. Traces of
iron, silica, and magnesium are
changed to minerals such as green
olivine or amber-colored garnet.

QUARTZITE

When sandstone rich in silica is changed by
the heat of contact metamorphism, it is
transformed into a rock called quartzite. It looks
a little like brown sugar but it is much tougher. In
fact, it is one of the toughest rocks around.
Because it is so tough and takes so long to
weather, it often pokes up in bare, jagged rock
outcrops almost devoid of vegetation. The more
it is compressed and
heated, the tougher it
becomes. Despite this, it is rarely used as a
building stone, simply because it is so hard to cut.
Some quartzites in Western Australia are among
the oldest rocks on Earth, some 3,500 million
years old.

MARBLE (above) *gets a
white, almost sugary look
as limestone is baked by the
heat of an igneous
intrusion.*

QUARTZITE (right) *is
one of the toughest of
all rocks, and many
samples have survived
since the very earliest era
of Earth's history.*

ID CHECK:
METAMORPHIC ROCKS

• **HORNFELS**
TYPE: *Contact metamorphism of shale*
GRAIN SIZE: *Fine*
TEXTURE: *Even*
COLOR: *Gray, blue-black*
CHARACTERISTICS: *Tends to splinter and
often contains large crystals*

• **MARBLE**
TYPE: *Contact metamorphism of
limestone*
GRAIN SIZE: *Medium–coarse*
TEXTURE: *Sugary*
COLOR: *White, streaky, variable*
CHARACTERISTICS: *Bright, pure colors
and shiny, sugar-like crystals*

• **QUARTZITE**
TYPE: *Contact metamorphism of
sandstone*
GRAIN SIZE: *Medium–coarse*
TEXTURE: *Sugary*
COLOR: *Brown, buff*
CHARACTERISTICS: *Very tough rock like
brown sugar*

Metamorphic rocks: regional

W HEN ROCK IS CRUSHED and baked beneath deep mountains that are thrown up between colliding continents, rocks may be metamorphosed over wide areas. Regional metamorphism can be low, medium or high grade, depending on just how severe the conditions are. In many places, the pressures can be so intense that all crystals line up in the same direction, creating a distinct stripy look, called schistosity.

SLATE (above) With its smooth, dark-gray plate-like structure, slate is one of the most easily recognized of all metamorphic rocks.

SCHIST (below) is a large family of metamorphosed rock. What all schists have in common is distinct layering, called schistosity, which can sometimes be confused with the layers in sedimentary rock.

SLATE

Slate is a distinctive, brittle, smooth gray rock that flakes into sheets. It is created by low-grade regional metamorphism of mudstone and shale, away from the focus of most intense metamorphism. Although the temperature here is fairly low, the pressure is enough to align mica and chlorite minerals into smooth, flat sheets. It splits or cleaves very easily and some of these sheets can be huge.

Slate may be very brittle, but it is actually quite weather resistant, which is why it is often seen in craggy outcrops in mountain regions. The combination of weather resistance and the ease with which it breaks into flat sheets made slate the perfect roofing material for centuries. Until recently, it was used for blackboards. Now it is used for snooker and pool tables.

SCHIST AND PHYLLITE

The resilient rock that provides a firm base for the huge skyscrapers of Manhattan is schist, created by medium-grade regional metamorphism of mudstone,

MANHATTAN (below) The towering skyscrapers of New York's Manhattan are built on very firm foundations. Underlaying the entire island is a thick bed of tough schist.

- **SLATE**

TYPE: *Regional metamorphism of shale*
GRAIN SIZE: *Fine*
TEXTURE: *Breaks in plates*
COLOR: *Gray, purply-black*
CHARACTERISTICS: *Breaks into very distinctive, dark-gray smooth plates*

- **SCHIST**

TYPE: *Regional metamorphism of slate and shale*
GRAIN SIZE: *Fine—medium*
TEXTURE: *Schistose, foliated*
COLOR: *Green, purply-black*
CHARACTERISTICS: *Comes in different varieties, but all with fine, foliated texture*

- **GNEISS**

TYPE: *High-grade regional metamorphism of slate, shale, and schist*
GRAIN SIZE: *Medium—coarse*
TEXTURE: *Banded, granular*
COLOR: *Gray, zebra stripes*
CHARACTERISTICS: *Often distinguished by contorted banding*

PHYLLITE *(above) is a light-colored schist made as the shale-like rock pelite is metamorphosed. Its name is Greek for "leaf," and it has thin layers like the leaves of a book.*

shale and clay. As temperature and pressure increase towards the centre of metamorphism, slate becomes even shinier and its layers begin to deform and recrystallize. First it changes into a rock called phyllite. When the temperature reaches over 750°F (400°C), crystals of chlorite and biotite mica begin to form and the rock changes to schist. The mica crystals grow in line with the pressure, creating a unique banded look (schistosity) at right angles to the pressure. Occasional crystals of almandine garnet may form too.

There are various different types and colors of schist, depending on the main minerals they contain. Green schists, for example, have a great deal of green chlorite. Often schist and gneiss are layered together over and over again as the rocks were folded and crushed. Such alternating bands are seen in many places in the Alleghenies and Connecticut.

GNEISS

The word gneiss – pronounced "nice" – comes from the Old High German word for "sparkling," and that is exactly what gneiss does. Under a microscope, it can be seen to be made from tiny, iridescent crystals which are created by the most intense metamorphism of all. Sometimes it is created by the metamorphism of granite. Sometimes gneiss is created as increasing temperatures – over 900°F (500°C) – turn schist minerals into staurolite, cordierite, and sillimanite, along with clots of feldspar and quartz. Between Vermont and the White Mountains of New Hampshire, there is a clear sequence of rocks changing from shale, to slate, to phyllite, schist and finally gneiss. Gneiss is perhaps the toughest rock of all, and large areas of Greenland are made from gneiss that has existed for billions of years. The oldest known rock in the world is an Acasta gneiss from Canada, which has been dated at over 3,900 million years old.

GNEISS *(right) often has distinct banding, called joliation, which makes it very easy to recognize. This banding is usually at right angles to the pressure which altered the rock, and is caused by the way crystals realign under pressure.*

MINERAL CRYSTALS
Native elements

MOST MINERALS ARE made from combinations of chemicals but a few are native elements – that is, chemical elements occurring naturally by themselves, uncombined with any others. These native elements originate in igneous and metamorphic rock. But more durable elements, such as gold and diamond, may survive long after the rock has been broken up by the weather. They may then be washed into streams, where they can be discovered relatively easily.

METALS

GOLD

Gold's glittering, imperishable beauty has made it prized since the dawn of history. It has become the very symbol of wealth, and the Gold Standard is the standard by which all money is valued. Because it will rarely form compounds with water, oxygen, or any other element, it occurs naturally in pure form — and remains shiny and untarnished for thousands of years. Typically, it forms in cubic or octahedral crystals in hydrothermal veins in igneous rocks, associated with quartz and sulfide minerals such as stibnite.

It may also be found as small grains in rivers where it has been washed. Large lumps or "nuggets" are rare indeed. The famous 156-pound "Welcome Stranger" nugget found in Moliagul, Australia, in 1869, was very exceptional. Most of the great "gold rushes" of the past, such as the Klondike and the California rush, were spurred by the discovery of gold grains in river beds. Nowadays, most gold is mined from veins.

SILVER (below) is one of the few chemical elements to occur naturally. But wherever it is exposed to air, the surface quickly tarnishes black, so it is hard to spot.

SILVER

In Ancient Egypt, silver was called white gold and was valued even more highly than gold itself. There are silver mines in eastern Anatolia (Turkey) which were dug by the pre-Hittite people of Cappaddocia over 5,000 years ago. When polished, silver is a beautiful, shiny, white metal, but it quickly tarnishes with a black coating of silver sulfide. This is why it is hard to spot in nature, even though the branching, dendritic (tree-like) crystals can be quite distinctive. Like gold, it often forms in hydrothermal veins, typically in association with copper or

GOLD AND PLATINUM *(left) have long been amongst the most valuable metals, made into jewelry and prized for their beauty and durability. Unlike most other metals, gold never tarnishes or corrodes.*

galena but, unlike gold, it rarely forms nuggets, so is rarely seen in placer deposits.

PLATINUM

Platinum is a silvery metal even rarer and more precious than gold. It used to be found in placer deposits, often with gold. Now it is more likely to be found as tiny grains in ultrabasic igneous rocks like peridotite and gabbro. Platinum is easy to distinguish from other metals because although it is often magnetic like iron, there is no other metal quite so dense and so soft. The prime sources in North America are Sudbury, Ontario, and Alaska.

COPPER

Highly valued for its electrical conductivity, copper is one of the most distinctive of all metals. Most copper used today is extracted from its ore, chalcopyrite, but it can be found in pure native form. Pure copper is often found in sulfide veins in warm desert areas, and in cavities in ancient lava flows. Like silver, copper crystals grow in branching dendritic or wiry masses, made of tiny cubic crystals. Typically, bright green and turquoise stains of malachite and azurite known as "copper blooms" on the rocks will alert you to the presence of copper. To see the copper, you have to scrape this stain away.

ID CHECK:
NATIVE ELEMENTS

• **GOLD**
CRYSTAL SYSTEM: Cubic
HARDNESS: 2.5–3
STREAK: Golden-yellow
CLEAVAGE: None
LUSTER: Metallic
HABIT: Dendritic, grains
MAIN CHEMICALS: Gold

• **SILVER**
CRYSTAL SYSTEM: Cubic
HARDNESS: 2.5–3
STREAK: Silver-white
CLEAVAGE: None
LUSTER: Metallic
HABIT: Dendritic, massive
MAIN CHEMICALS: Silver

• **COPPER**
CRYSTAL SYSTEM: Cubic
HARDNESS: 2.5–3
STREAK: Copper red
CLEAVAGE: None
LUSTER: Metallic
HABIT: Dendritic, cubic
MAIN CHEMICALS: Copper

BRONZE BUDDHA *(left) Copper was one of the first metals to be used by man, alloyed with tin in prehistoric times to make bronze, which can be cast into all kinds of shapes.*

COPPER *(right) is sometimes found in pure form in the ground — but is usually covered with a green stain of malachite or azurite.*

Non-metals

SULFUR

Sulfur is also known as brimstone or "burning stone," and it does indeed burn, with a blue flame, releasing choking fumes of sulfur dioxide and melting away into a dark-yellow liquid. Bright-yellow sulfur crystals typically form around fumaroles-volcanic chimneys which belch sulfurous fumes – and they are found in abundance in the Valley of Ten Thousand Smokes in Alaska. Most of the world's sulfur comes from beds of limestone and gypsum in Poland, Sicily, and the Gulf of Mexico.

SULFUR *(above) is one of the easiest minerals to recognize, forming bright-yellow crystals around the chimneys of volcanoes and hot springs.*

DIAMOND

Diamond is the world's hardest natural substance, getting its name from the Ancient Greek word *adamantos* meaning "invincible." It is pure carbon, like coal, but it looks like glittering glass because the carbon has been transformed by enormous pressure. It is now possible to make diamonds artificially by squeezing carbon under such pressures. But in nature such pressure is rare, which is one reason why diamonds are so precious. Most diamonds found today are very old indeed, and formed deep in the Earth billions of years ago, as the world's tectonic plates crunched together.

GRAPHITE

Like diamond, graphite (once known as plumbago) is pure carbon but it could hardly be more different. Graphite is dull, dark gray and very soft. It typically occurs as scaly masses where limestones rich in plant remains have been metamorphosed into marble.

DIAMOND *(above) is the hardest of all natural substances, formed under huge pressures deep in the Earth, and then brought to the surface by volcanic activity.*

Halides

The best-known halide is ordinary salt, like table salt. But all halides are salts, in which a metallic element is combined with one of the elements called halogens — chlorine, bromine, fluorine and iodine. They all dissolve easily in water, so many occur only under special conditions. Only common salt is easy to find, despite its solubility, simply because it is so abundant.

GRAPHITE AND PENCILS *(left)*
Like diamond, graphite is pure carbon. But it is very different — so soft, powdery, and black that it is used as the lead in pencils. It is typically found in association with marble and other metamorphic rocks.

HALITE

Halite is common salt or rock salt – the same salt you use on food. You can make your own crystals by evaporating sea water. Most of the world's salt deposits were formed by the evaporation of sea water. Look through a magnifying glass at table salt and you will see that the crystals are cubic.

FLUORITE

Fluorite comes in all kinds of forms and brilliant colors. It is only because they are so soft and soluble that they are not used for jewelry. Some fluorite crystals are bright pink, like Turkish delight.

HALITE AND SALT (above) The salt you use in food comes either from the sea or from the ground. Wherever it comes from, it is the mineral halite. Solid halite forms as sea water evaporates, leaving salt behind.

ATACAMITE

Getting its name from the Atacama desert in Chile where the best samples are found, dark-green atacamite often forms in association with bright-green malachite as copper sulfide minerals are altered by exposure to air. It is quite rare, and was once highly prized as an attractive sand for drying ink before blotting paper was invented.

FLUORITE (above) typically forms yellowish, cube-shaped, clear crystals which often glow fluorescently. But the crystals can be many other colors, including pink.

ATACAMITE (right) is a rare, though not especially precious, mineral that typically forms in deserts — mostly in Chile and South Australia — in association with copper ore.

ID CHECK:
NATIVE ELEMENTS & HALIDES

• SULPHUR
CRYSTAL SYSTEM: Orthorhombic
HARDNESS: 1.5–2.5
STREAK: Pale yellow
CLEAVAGE: None
LUSTER: Resinous
HABIT: Tabular, crusts
MAIN CHEMICALS: Sulfur

• DIAMOND
CRYSTAL SYSTEM: Cubic
HARDNESS: 10
STREAK: None
CLEAVAGE: Perfect
LUSTER: Adamantine
HABIT: Octahedral crystals
MAIN CHEMICALS: Carbon

• HALITE
CRYSTAL SYSTEM: Cubic
HARDNESS: 2.5
STREAK: White
CLEAVAGE: Perfect
LUSTER: Vitreous
HABIT: Cubic crystals, massive
MAIN CHEMICALS: Sodium chloride

Sulfides and sulfosalts

SULFIDES ARE COMPOUNDS of sulfur, usually with a metal, which is why they often have a metallic luster and include some of the world's most important metal ores, including lead, zinc, iron, and copper. They tend to be brittle and heavy. Typically, they form in hydrothermal veins, where hot fluids bring minerals up through cracks in the ground. Sulfosalts are compounds of sulfur, a metal, and a semi-metal such as arsenic or antimony.

GALENA

Galena, or "lead-glance," is the main ore of lead, and crystals form in dark-gray cubic masses. The Joplin District of Missouri, Kansas, and Oklahoma is known for its large galena crystals, but the main mining district in the United States is the Tri-state in Mississippi. Sometimes, the lead may be associated with tiny amounts of silver, and galena is a main source of silver as well as lead.

GALENA *(above) is lead ore, and forms in attractive, shiny, dark-silvery colored, cube-shaped crystals — typically in veins or around small igneous intrusions.*

CINNABAR

Usually a distinctive bright or brick red, cinnabar is the main ore of mercury. It crystallizes in hydrothermal veins in sedimentary rocks only when the temperature has dropped quite low, so it is typically found fairly near the surface or around hot springs. The best deposits in the United States are in California.

SPHALERITE

Sphalerite is the main ore of zinc, and its crystals are dodecahedral. It is typically black, and has the nickname "black jack." But it can also come in other colors including red ("ruby jack"), green, and brown, so is not always easy to recognize. It usually forms in hydrothermal veins with galena and pyrite. It is very widespread, and the Joplin District of Missouri, Kansas, and Oklahoma is one of the places where it is found in North America.

KEEPING SULFIDES BRIGHT

Most sulfides such as pyrite and marcasite quickly tarnish when exposed to the air. There is no easy solution to this problem, but wrapping samples in plastic wrap and storing them in a dark, dry place will delay the deterioration. Coating with varnish also helps.

CINNABAR *(above right), the ore of mercury, is one of the most distinctive of all minerals — a bright or brick red. Wash your hands after handling this.*

SPHALERITE *(left) is the main ore of zinc. It typically forms pyramid-shaped crystals — but they can be many colors, including amber (honey blonde), ruby, white, and black.*

CHALCOPYRITE

Forming in hydrothermal veins and pegmatites along with galena, pyrite, and sphalerite, chalcopyrite is the main ore of copper. Its surface is often coated with a dark-greeny, purply, slightly iridescent tarnish which earns it the name "peacock copper." Beneath the coating, the pyramidal crystals are a yellow gold, like pyrite, only yellower, and more easily scratched. Chalcopyrites called "porphyry coppers" are mined in Bingham, Utah.

BORNITE

Bornite is another important copper ore, and is the true "peacock ore" with a purply iridescent tarnish similar to chalcopyrite. Indeed, the two can be hard to tell apart. Crystals are fairly rare, and it is usually found as masses mixed in with other minerals such as chalcopyrite, as it is in Arizona.

CHALCOPYRITE *(above) is the main ore of copper, and is usually coated in a greenish-colored tarnish, as copper combines with oxygen in the air.*

BORNITE *(right) is another important ore of copper. It can look very like chalcopyrite, but bornite is more likely to have a purplish "peacock" sheen.*

ID CHECK:
SULPHIDES AND SULPHOSALTS ROCKS

• **GALENA**
CRYSTAL SYSTEM: Cubic
HARDNESS: 2.5
STREAK: Lead gray
CLEAVAGE: Perfect
LUSTER: Metallic
HABIT: Cubes, granular
MAIN CHEMICALS: Lead sulfide

• **CINNABAR**
CRYSTAL SYSTEM: Trigonal
HARDNESS: 2.5
STREAK: Red
CLEAVAGE: Perfect
LUSTER: Adamantine
HABIT: Small, granular masses
MAIN CHEMICALS: Mercuric sulfide

• **SPHALERITE**
CRYSTAL SYSTEM: Cubic
HARDNESS: 3.5
STREAK: White, yellow-brown
CLEAVAGE: Perfect
LUSTER: Adamantine
HABIT: Cubes, fibrous
MAIN CHEMICALS: Zinc sulfide

• **CHALCOPYRITE**
CRYSTAL SYSTEM: Tetragonal
HARDNESS: 3.5—4
STREAK: Greenish-black
CLEAVAGE: Imperfect
LUSTER: Metallic
HABIT: Massive, twinning
MAIN CHEMICALS: Copper, iron sulfide

PYRITE

Sometimes known as "fool's gold" because people have often mistaken it for gold, pyrite, or iron sulfide, is one of the most common of all the sulfides, found in almost all kinds of rocks. Indeed, any rock that looks a little rusty probably contains pyrite. Crystals can be cubic or octahedral. When struck against a metal object, it gives off sparks, which is why it has been used to light fires since prehistoric times.

ORPIMENT AND REALGAR

Richly colored crystals like crystallized orange make orpiment unmistakable. Realgar is bright red, like cinnabar, but chunks of it set in carvings by the Ancient Chinese have almost disintegrated through long exposure to the air. Both orpiment and realgar are sulfides of arsenic, and highly poisonous. The best examples are found at the Getchell gold mine in Nevada.

MARCASITE

Marcasite looks very like pyrite, but forms at lower temperatures and is often found in limestones and chalk. Sometimes it forms in long crests called cockscombs because they look like the comb of a rooster. It also forms in nobbly nodules that split open to reveal a radiating pattern of crystals. People have often mistaken these nodules for meteorites.

PYRITE *(above) can look so like gold that people have been fooled into thinking that they have made their fortune — but it is simply iron sulfide.*

ORPIMENT *(left) is beautiful but deadly. Its orange crystals contain arsenic and are highly poisonous.*

MARCASITE *(above) forms in nodules that can look strange enough to have come from outer space; many people have mistaken them for meteorites.*

REALGAR *(right) is, like orpiment, attractive but lethal. Its bright-red crystals similarly contain arsenic.*

Sulfates and others

THE SULFATES ARE a large and widespread group of minerals — soft, light colored, and typically translucent. They form when one or more metals combines with a sulfate, which is a partnership of sulfur and oxygen.

GYPSUM *(above) is typically a soft, white mineral — the basis for various kinds of plaster.*

GYPSUM AND SELENITE
Soft and white, gypsum is a very common mineral that forms in thick beds where salty water evaporates. In many places, gypsum is massive and powdery. But in the desert, crystals grow around grains of sand into flower-like shapes called desert roses. When gypsum is heated, it loses some of its water content and becomes a powder known as plaster of Paris.

DESERT ROSE *(above) forms as gypsum grows round grains of sand in the desert.*

BARITE
Barite is useful because it is so dull and neutral. It is used as the base for white paint and to soak up dangerous rays from hospital X-rays.

PHOSPHATES, VANADATES, ARSENATES
Phosphates, vanadates, and arsenates are usually secondary minerals that form as ore minerals are broken down by the weather. Combining with metals can give them vivid colors, such as bright green olivenite or blue-green turquoise.

TUNGSTATES, MOLYBDATES, AND URANATES
This is a small group of minerals as colorful and interesting as the phosphates — but good crystals are hard to find.

BARITE *(above) may look dull, but it is actually one of the most useful minerals, used as the basis for white paint.*

APATITE *(left) typically forms very clear hexagonal column crystals. When it is tinged green, it is called asparagus stone.*

TURQUOISE *(right) is one of the most beautiful minerals, prized by the Ancient Egyptians and the Aztecs.*

ID CHECK:
SULPHIDES AND SULPHOSALTS ROCKS

- **PYRITE**
  CRYSTAL SYSTEM: Cubic
  HARDNESS: 6.5
  STREAK: Green-black
  CLEAVAGE: Poor
  LUSTER: Metallic
  HABIT: Cubes, granular
  MAIN CHEMICALS: Iron sulfide

- **MARCASITE**
  CRYSTAL SYSTEM: Orthorhombic
  HARDNESS: 6.5
  STREAK: Gray-brownish
  CLEAVAGE: Poor
  LUSTER: Metallic
  HABIT: Tabular, radiating masses
  MAIN CHEMICALS: Iron sulfide

- **GYPSUM**
  CRYSTAL SYSTEM: Monoclinic
  HARDNESS: 2
  STREAK: Lozenge chips
  CLEAVAGE: Poor
  LUSTER: Vitreous
  HABIT: Fibrous, massive
  MAIN CHEMICALS: Calcium sulfate

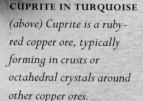

Oxides

OXIDES ARE ONE of the largest and most varied of all the groups of minerals. They are formed by the combination of a metal with oxygen, and the only metals which do not form oxides are silver and gold. They include everything from the commonest, dullest ores like bauxite (aluminium ore), to the rarest, most precious gems such as rubies and sapphires, and they vary in color from the rich red of cuprite to the dark black of magnetite.

Some form deep in the Earth's crust in magma, or in very hot veins. These "primary" oxides tend to be hard, like corundum, and are "anhydrous" which means they contain little water. "Secondary" oxides like bauxite form near the surface when minerals such as sulfides are attacked by the atmosphere. These secondary oxides tend to be soft and "hydrous" — that is, contain water — and they break down to form soil.

CUPRITE

Red as ruby, cuprite is the richest copper mineral of all (88 percent copper) and is formed as copper sulfide ores are exposed to the weather for a while. It seems to form best in dry, tropical soils where weathering reaches deep down into the ground, and some of the best examples come from Namibia in Africa.

CUPRITE IN TURQUOISE *(above) Cuprite is a ruby-red copper ore, typically forming in crusts or octahedral crystals around other copper ores.*

SPINEL

Spinels are a group of minerals, including magnetite and chromite, that are made from combinations of oxide minerals. But the best known of the spinels is the brilliant gem spinel, which comes in a wide range of colors, including red, blue, turquoise, and violet. Gem spinel typically forms in metamorphosed gneisses and marbles. Spinel minerals melt and recrystallize easily, so they are often used to make synthetic gems such as sapphires and aquamarines.

SPINEL *(above) is a gem in its own right, but is often melted down and used to make synthetic gems.*

HEMATITE *(left) is the major ore for iron. It forms in many shapes and colors, from petal-shaped "roses" to kidney-shaped "kidney ore" and blood-colored bloodstone.*

SAPPHIRE *(right) is the name for all gem corundums which are not red, but the best known are brilliant blue.*

HEMATITE

Hematite is the most important ore of iron. It forms in huge layers in sedimentary rocks and is quarried on a massive scale. It gets its name from the Ancient Greek word for "blood," and if soils or sedimentary rocks have a reddish hue, the color often comes from hematite. Hematite also forms in reniform (kidney-shaped) lumps, called kidney ore, and in pockets called "iron roses," as found near Saint Gotthard in the Swiss Alps.

CORUNDUM: RUBY AND SAPPHIRE

Only diamond is harder than corundum, which is why it is often used as an abrasive on "emery" paper and as a knife-sharpening stone. It can form as the precious gems ruby and sapphire — the best of which have long come from Burma (rubies) and Kashmir (sapphires).

RUBY *(above) is a dark-red gem, one of the many gem forms of corundum, colored red by traces of chromium oxide.*

CORUNDUM *(above) is one of the hardest of all minerals, only a little less hard than diamond. It is found in many gem varieties, including sapphire and ruby.*

KIDNEY ORE *(left) Occasionally, reddish hematite forms in bulbous reniform (kidney-shaped) nodules with a smooth surface. This is known as kidney ore.*

ID CHECK:
OXIDES

• **CUPRITE**
CRYSTAL SYSTEM: *Cubic*
HARDNESS: *3.5*
STREAK: *Brown-red*
CLEAVAGE: *Perfect*
LUSTER: *Metallic, dull*
HABIT: *Encrusting*
MAIN CHEMICALS: *Copper oxide*

• **HEMATITE**
CRYSTAL SYSTEM: *Trigonal*
HARDNESS: *6*
STREAK: *Dark red*
CLEAVAGE: *None*
LUSTER: *Metallic*
HABIT: *Massive, botyroidal masses*
MAIN CHEMICALS: *Iron oxide*

• **CORUNDUM**
CRYSTAL SYSTEM: *Trigonal*
HARDNESS: *9*
STREAK: *None*
CLEAVAGE: *None*
LUSTER: *Adamantine, vitreous*
HABIT: *Tabular, granular*
MAIN CHEMICALS: *Aluminium oxide*

Carbonates

CARBONATES ARE MINERALS that typically form when metals and semi-metals combine with a carbonate, which is a partnership of carbon and oxygen. Although some are brought to the surface by solutions of hot fluids from deep in the Earth's crust, most form by the alteration of minerals on the surface. The carbonates include bright-green malachite and deep blue azurite, but most are fairly soft, pale, and translucent.

CALCITE *(above) No other mineral comes in as many forms as calcite – over 100 different ones; it also comes in many hues.*

THE CALCITES: CALCITE, SIDERITE, MAGNESITE, AND RHODOCHROSITE

Calcite is the fur inside kettles and the limescale around the bath in areas of hard water. It is also the main ingredient in limestone, marble, and chalk, and develops into fantastic dripstone formations in limestone caves. The crystals grow in many different shapes and forms, including dogtooth spar, which looks like the fangs of a hungry dog. Most calcite is white, but it can be tinted almost any color by impurities. Perhaps the most distinctive variety of calcite crystal is Iceland spar, which forms in beautifully clear, flat, rhombohedral crystals that look like shards of ice.

Besides calcite itself, there are a range of other calcites, including magnesite, siderite and rhodochrosite. All of them have rhombohedral crystals. Siderite is an important ore of iron. Dull, white magnesite forms when serpentine is altered by hot water, creating white veins which are mined for magnesium. Rhodochrosite is rose-red calcite, often used as a gem.

SIDERITE *(above) is chemically iron carbonate and is also known as iron spar, an important ore of iron.*

THE ARAGONITES: ARAGONITE, CERUSSITE, AND WITHERITE

Aragonites have orthorhombic crystals, and often grow in stars made of three crystals. Aragonite itself is the pearly mineral that shimmers iridescently on the inside of some sea shells. But it typically forms in crusts around hot springs, and occasionally forms on cave walls in fantastic little bunches called *flos ferri* or "iron flowers." Cerussite has a vitreous luster almost like gelatine and is an important ore of lead. Witherite is the rare carbonate that forms when the calcium of calcium carbonate is replaced by barium.

MAGNESITE *(above) is a rare carbonate known as bitter spar, typically found in areas of dolomite limestone.*

RHODOCHROSITE *(right), or raspberry spar, is typically a delicate rose red and forms in balls called druses.*

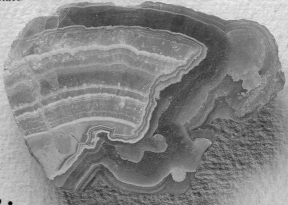
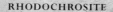

ARAGONITE *(left) varies considerably in color and form — from a dull brown to the brilliant pearly sheen found on the inside of oyster shells.*

MALACHITE AND AZURITE

Unlike the other carbonates, malachite and azurite are both vividly colored. Malachite (copper carbonate) is the bright-green tarnish on copper, and usually forms as other copper minerals are altered by weathering. Very hard crusts of malachite, like those mined nowadays in Zimbabwe's copper belts, were used for icons in Russian churches.

Azurite, like malachite, is formed by the weathering of other copper minerals. But the presence of water in the crystal helps turn it bright blue rather than green. It was this brilliant blue that made it popular with painters in the Renaissance as a pigment. Both azurite and malachite are not only major copper ores, but colorful signposts to finding other copper minerals.

ID CHECK:
CARBONATES

• CALCITE
CRYSTAL SYSTEM: Trigonal
HARDNESS: 2.5–3
STREAK: White
CLEAVAGE: Perfect rhomb
LUSTER: Vitreous
HABIT: Various
MAIN CHEMICALS: Calcium carbonate

• ARAGONITE
CRYSTAL SYSTEM: Orthorhombic
HARDNESS: 3.5–4
STREAK: White
CLEAVAGE: Distinct
LUSTER: Vitreous
HABIT: Many
MAIN CHEMICALS: Calcium carbonate

CERUSSITE *(above left) is white lead ore and forms where galena, another lead ore, is exposed to the weather.*

WITHERITE *(above right) is barium carbonate — a rare and highly poisonous mineral.*

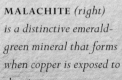

MALACHITE *(right) is a distinctive emerald-green mineral that forms when copper is exposed to the air.*

AZURITE *(below) is instantly recognizable from its deep-blue color. It typically forms where copper ore is exposed to the air and oxidizes.*

Silicates

QUARTZ *(right) is the most common of minerals and comes in a huge variety of forms, but most are glassy brown or white.*

ORTHOCLASE FELDSPAR *(above) is the basic mineral in most intrusive igneous rocks, but rarely forms large, individual crystals.*

PLAGIOCLASE FELDSPAR *(above) is calcium-sodium feldspar, while orthoclase is basically potassium Albite; anorthite and many others are plagioclases.*

SILICATES ARE THE most common group of minerals by a long way. There are more silicates than all the rest of the minerals put together, and a few silicates like quartz and the feldspars are so abundant that they make up a huge proportion of igneous and metamorphic rocks. Silicates are essentially metals combined with silicon and oxygen. Because silicon and oxygen are the most abundant of all the elements in the Earth's crust, the most common kind of magma bubbling up towards the surface is silicate magma. The bulk of silicate minerals form as this solidifies. Silicates do form in other ways, such as during intense metamorphosis, but most form from magma.

QUARTZ

Quartz (silicon dioxide) is the single most common mineral in the Earth's crust, a very hard mineral found in all but a few very basic rocks. The shiny, gray crystals in granites are quartz. Pebbles on beaches are usually quartz. When silicon dioxide crystallizes at low temperatures in volcanic cavities, the crystals can be so tiny that the mineral looks more like porcelain than crystal. This "microcrystalline" quartz or chalcedony often has a dull or waxy look, but comes in such stunning colors that it is often sought after for gems – including blood-red carnelian, wine-red jasper, and apple-green chrysopase.

FELDSPARS AND FELDSPATHOIDS

Together, the two main types of feldspar – orthoclase and plagioclase – are even more abundant than quartz. Orthoclase feldspar is one of the main ingredients of granite and many other igneous rocks, as well as metamorphic rocks like gneiss and sediments like arkose. Feldspars vary in color from milky white and pink to very dark green with minute traces of

OPAL *(left) is a beautiful stone but rarely forms crystals. Instead, it comes in shapeless masses – but gets its beauty from its rainbow sheen, called opalescence.*

different chemicals. Occasionally, in veins and volcanic cavities, feldspars form into large crystal gems such as amazonstone and labradorite.

MICA AND MUSCOVITE

Mica and muscovite form when potassium and aluminium link up to silicates. Biotite mica is one of the main ingredients in granites, gneisses, and schists. Its dark brown or black, flaky crystals look rather like the pages of a book. Muscovite mica is the clear mineral that gives schist its distinctive silky sheen. It gets its name from Muscovy, the old name for Russia, where it was once used for windows instead of glass. It is still used for stove windows because it is so resistant to heat.

GARNETS

Garnets are a group of gems that form at very high temperatures – during metamorphosis, for instance. They are often found in gneisses, schists, and marbles, as well as peridotites that have come up from the Earth's hot interior. You may find them in river beds, too, because they are so tough they survive long after the rock around them has been worn away. They include red almandine, ruby-red pyrope, green grossularite, and green andradite.

ID CHECK: SILICATES

• **QUARTZ**
CRYSTAL SYSTEM: *Trigonal*
HARDNESS: 7
STREAK: *White*
CLEAVAGE: *None*
LUSTER: *Vitreous, greasy*
HABIT: *Many*
MAIN CHEMICALS: *Silicon dioxide*

• **ORTHOCLASE FELDSPAR**
CRYSTAL SYSTEM: *monoclinic*
HARDNESS: 6
STREAK: *White*
CLEAVAGE: *Perfect*
LUSTER: *Vitreous, pearly*
HABIT: *Various, compact*
MAIN CHEMICALS: *Potassium*

• **PLAGIOCLASE FELDSPAR**
CRYSTAL SYSTEM: *Triclinic*
HARDNESS: 6
STREAK: *White*
CLEAVAGE: *Perfect*
LUSTER: *Vitreous, pearly*
HABIT: *Various*
MAIN CHEMICALS: *Sodium*

GARNET *(below) is a rare, usually wine-red gem that forms in metamorphic rocks as they are altered by heat and pressure.*

BIOTITE MICRA *(right), like all micas, is shiny because it breaks so cleanly. Biotite can be shiny black, dark brown, or green.*

GLOSSARY

ASTHENOSPHERE
The softer layer of the Earth immediately below its rigid outer shell (lithosphere), extending from 60 to 200 miles down.

APATITE
A common phosphate mineral found in rocks that provides the main source of phosphorus for plants when the rocks break down.

BATHOLITH
A huge igneous intrusion beneath the ground in a very rough dome shape, typically cooling to form granite rocks.

BED
A layer of sedimentary rock—the top and bottom of the bed mark a break as the sediments that formed the rock were laid down—that may have been due simply to a change in the season or to something more dramatic.

BEDDING PLANE
The break in the rock layers at the top and bottom of each rock bed.

DISCONTINUITY
A distinct break in the sequence of layers of rock.

DRIFT
Large deposits of loose material covering the surface, left behind by ice sheets, flooding rivers, or the wind.

DIKE
A kind of igneous intrusion. Dikes are thin, slanting or vertical sheets of volcanic rock injected into the rock, breaking across existing rock structures.

FAULT
A fracture along which blocks of rock slide past each other.

FOLD
A crumpling of the rock strata of the Earth's crust, typically caused as two tectonic plates collide.

FOLIATION
A banded structure within rock, usually caused by extreme heat and pressure.

INDEX FOSSIL
A particular type of fossil useful for dating rock strata.

INTRUSION
Where magma wells up into the ground but does not break the surface as in a volcano.

JOINT
A slanting or vertical crack in sedimentary rock beds, typically created as the rock dried out, or as the pressure of overlying rock was released.

LAMELLAR
Made of very thin, flattish layers.

LAVA
The molten rock that gushes out of volcanoes.

LITHOSPHERE
The Earth's rigid shell.

LOPOLITH
A saucer-shaped igneous intrusion formed where magma runs into a downfold in the rock structure.

MAGMA
The molten volcanic rock that wells up from the Earth's interior.

PHACOLITH
A small, lens-shaped igneous intrusion, formed where magma oozes between rock layers.

SCHISTOSITY
A fine banding of the crystals in metamorphic rock.

SERPENTINE
A greasy-looking silicate rock, also a mineral, formed when other minerals are altered by water.

SILL
A kind of igneous intrusion. Sills are thin, slanting or horizontal sheets of volcanic rock that follow existing rock structures, such as beds.

STOCK
Small-scale, drum-shaped igneous intrusions, typically offshoots of batholiths.

STRATA
Layers of sedimentary rock.

SUBDUCTION
When two tectonic plates collide and one is pushed down into the interior beneath the other, setting off volcanoes and earthquakes.

TECTONIC PLATES
The twenty or so giant slabs of rock that make up the lithosphere. The slow movement of these has a tremendous effect on the rocks of the Earth's surface, not only allowing magma to ooze to the surface to form igneous rocks, but twisting and distorting rock structures to throw up mountains, crumple rock layers, and create new metamorphic rocks by subjecting them to huge pressures.

TRAVERTINE
A limey rock formed from mineral deposits around springs, named after the Italian city of Travertine.

INDEX

Page numbers in *italics* refer to captions/illustrations.

CREDITS

Quarto would like to acknowledge and thank the following for use of pictures on the pages below:

Clark/Clinch: 8b, 25, 49t; e.t. archive: 34l; Image bank: 12c, 13t, 16cl, 26br, 36, 39l, 40, 43, 45r, 51t, 52b, 53t, 57t, 60b; Chris Pellant: 6b, 7t, 9bl, 11 t & b, 17t, 27t; Pictor: 58cl; Vaughan Fleming: 26tl, 46.

Key   t - top   b - bottom   r - right   l - left   c - centre

All other pictures are the copyright of Quarto

Quarto would like to thank the following for supplying us with rock samples for photography:

R. Holt and Company Ltd
98 Hatton Garden
London EC1N 8NX

Richard Taylor Minerals
20 Burstead Close
Cobham
Surrey
KT11 2NL

Special thanks also go to **United States Geological Survey** for providing us with a geologic map of the Grand Canyon and to **The Geological Association of London** and **Gregory, Botley & Lloyd** for supplying us with props and materials for photography.